江苏大学英文教材基金资助出版

ELECTROTECHNICS

电工学

主编

陈　山

副主编

朱　莉　盛碧琦　陈　娇

江苏大学出版社
JIANGSU UNIVERSITY PRESS
镇　江

图书在版编目(CIP)数据

电工学＝Electrotechnics：英文 / 陈山主编. —镇江：江苏大学出版社，2018.8
ISBN 978-7-5684-0857-8

Ⅰ.①电… Ⅱ.①陈… Ⅲ.①电工－高等学校－教材－英文 Ⅳ.①TM

中国版本图书馆 CIP 数据核字(2018)第 190231 号

电工学
Electrotechnics

主　　编/陈　山
责任编辑/李经晶
出版发行/江苏大学出版社
地　　址/江苏省镇江市梦溪园巷 30 号(邮编：212003)
电　　话/0511-84446464(传真)
网　　址/http://press.ujs.edu.cn
排　　版/镇江文苑制版印刷有限责任公司
印　　刷/虎彩印艺股份有限公司
开　　本/787 mm×1 092 mm　1/16
印　　张/13.25
字　　数/440 千字
版　　次/2018 年 8 月第 1 版　2018 年 8 月第 1 次印刷
书　　号/ISBN 978-7-5684-0857-8
定　　价/45.00 元

PREFACE

Since the reform and opening up, China has entered a period of rapid economic development, quickly integrated into the tide of world economic development. The economic globalization and cultural diversity have made the internationalization become the universities' top priority. Now, more and more international students study in China, and they want to learn Chinese culture and Chinese characters. Some of the students coming from African and Arabic countries want to learn science and engineering. Because of different education backgrounds, these students learned a little knowledge about physics and mathematics. Even we have many different photocopy textbooks, but they are too profound and expensive to those students. Electrotecnics is a series of books, including *Electrical*, *Electronics*, *Experiment Guiding Book* and *Problems of Electrotecnics*. As an Electrical book this book retains the basic laws and theorems, advanced techniques of network analysis, transient circuit, sinusoidal steady-state analysis, transformers and motors. This book also uses Multisim to simulate the circuits in every part, and will help students to understand the experiments that they do in the lab. This book is designed for a one-to two-semester course in electrical circuits and linear circuit analysis, and the presentation is geared to readers who are being exposed to the basic concepts of electric circuits for the first time.

Electrical Engineer(EE) is a fundamental technical course for non-EE-major undergraduates. The learning quality of this course will affect the students' learning of subsequent courses. The current climate of internationalization of undergraduate education has also brought new requirements for these courses.

This book also can be used as the English textbook of the fundamental course for those non-EE-major undergraduates, and it can serve as a reference book for the technicians in the electrical engineering and automation fields as well.

As for the ability of the author is limited, the contents of the book are insufficient. Hope the readers and some teachers could propose some good advice in order to make some revisions in the next edition.

FEATURES

The book begins with the development of electric circuit elements and electric variables used to describe them. Then resistive circuits are studied in Chapters 1—3 to provide an in-depth introduction to the concept of a circuit and its analysis. Many useful theorems and principles developed with insightful analysis of electric circuits are considered.

Chapter 4 begins the study of energy storage elements—inductor and the capacitor and time response of electric circuits that incorporate energy storage elements.

Chapter 5 begins with the characters of a sinusoidal signal. Then computations of sinusoids by means of rectors are introduced. Concepts and connections of complex impedance are also considered. Some general analytical techniques of AC circuits are considered in the remainder of this chapter.

Power calculations for general loads in AC circuits are considered in Chapter 6.

Chapter 7 begins with the balanced three-phase source. The analysis for Wye-and Delta-connected loads and the power calculations are also covered in this chapter.

The basic concept of magnetic field and the law of magnetic circuit are introduced in chapter 8 and it serves as the foundation to the introduction of transformers. Then, further study of magnetic materials, various types of transformer and the corresponding calculation are followed.

Chapter 9 begins with the principles applied to rotating machines. Three-phase induction motor is emphasized, including the principles of operation and the torque-speed characteristics.

CONTENTS

Answers of exercises in this book can be checked on http://press.ujs.edu.cn/NewsView.Asp? ID = 491&SortID = 64 or by scaning the QR code.

CHAPTER 1
CIRCUIT ELEMENTS

Introduction

This chapter presents the fundamental laws that govern the behavior of electric circuits, and it serves as the foundation to the remainder of this book. The chapter begins with a series of definitions to acquaint the reader with electric circuits; then, the two fundamental laws of circuit analysis are introduced: Kirchhoff's current and voltage laws. With the aid of these tools, the concepts of electric power are presented. Following these preliminary topics, the emphasis moves to basic analysis techniques—voltage and current dividers, and to some application examples related to engineering use of these concepts.

1.1 Voltage and current

The concept of electric charge is the basis for describing all electrical phenomena. Let's review some important characteristics of electric charge.

1.1.1 The charge

The **charge** is bipolar, meaning that electrical effects are described in terms of positive and negative charges.

The charge exists in discrete quantities, which are integral multiples of the electronic charge,

$$1.6022\times10^{-19}\,\text{C}.$$

Electrical effects are attributed to both the separation of charges and charges in motion.

In circuit theory, the separation of charges creates an electric force (voltage), and the motion of charges creates an electric fluid (current).

1.1.2 Electrical current

1.1.2.1 Current

Electrical **current** is the time rate of flow of electrical charges through a conductor or circuit element. The units are amperes (A), which are equivalent to coulombs per second (C/s).

Conceptually, to find the current for a given circuit element, we first select a cross section of the circuit element roughly perpendicular to the flow of current. Then, we select a reference direction of flow. Thus, the reference direction points from one side of the cross to the other side. This is illustrated in Figure 1.1.1.

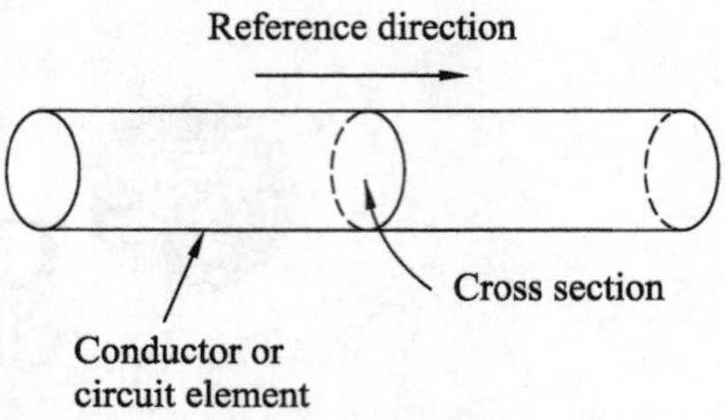

Figure 1. 1. 1 Current is the time rate of charge flow through a cross section of a conductor or circuit element

Suppose that we keep a record of the net charge flow through the cross section. Positive charge crossing in the reference direction is counted as a positive contribution to net charge. Positive charge crossing opposite to the reference direction is counted as a negative contribution. Furthermore, negative charge crossing in the reference direction is counted as a negative contribution, and negative charge against the reference direction is a positive contribution to charge.

Thus, in concept, we obtain a record of the net charge in coulombs as a function of time in seconds denoted as $q(t)$. The electrical current flowing through the element in the reference direction is given by

$$i(t)=\frac{dq(t)}{dt} \tag{1.1.1}$$

A constant current of one ampere means that one coulomb of charge passes through the cross section each second.

1. 1. 2. 2 Reference direction

In analyzing electrical circuits, we may not initially know the actual direction of current flow in a particular circuit element. Therefore, we start by assigning current variables and arbitrarily selecting a reference direction for each current of interest. It is customary to use the letter i for currents and distinguish different currents. This is illustrated by the example in Figure 1. 1. 2, in which the boxes labeled A, B, and so on represent circuit elements.

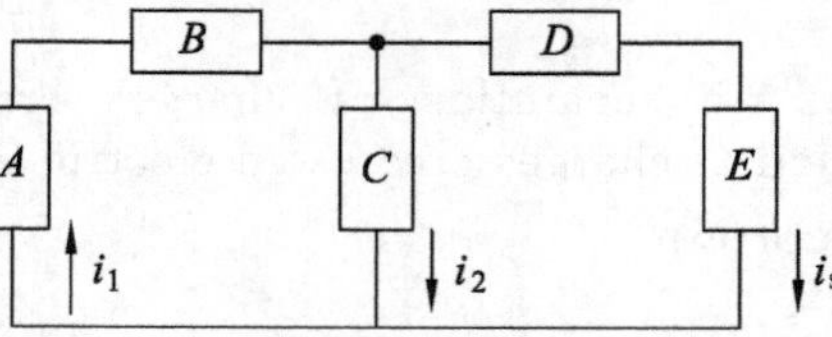

Figure 1. 1. 2 In analyzing circuits, we frequently start by assigning current variables i_1, i_2, i_3, and so forth

After we solve for the current values, we may find that some currents have negative values. For example, suppose that $i_1=-2$ A in the circuit of Figure 1. 1. 2, because i_1 has a negative value, we know that current actually flows in the direction opposite to the reference initially selected for i_1. Thus, the actual current is 2 A flowing downward through element A.

1. 1. 2. 3 Direct current and alternating current

When a current is constant with time, we say that we have **direct current**, abbreviated as DC. On the other hand, a current that varies with time, reversing direction periodically, is called alternating current, abbreviated as AC. Figure 1.1.3 shows the values of a DC current and a sinusoidal AC current versus time.

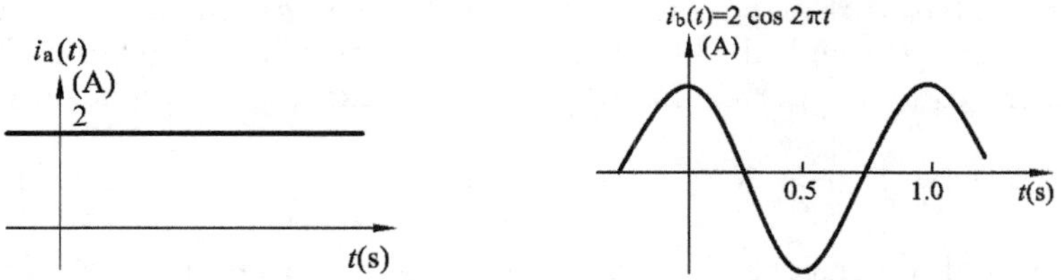

(a) DC current (b) AC current

Figure 1.1.3 Examples of DC and AC currents versus time

When $i_b(t)$ takes a negative value, the actual current direction is opposite to the reference direction for $i_b(t)$. The designation AC is used for other types of time-varying currents, such as the triangular and square waveforms shown in Figure 1.1.4.

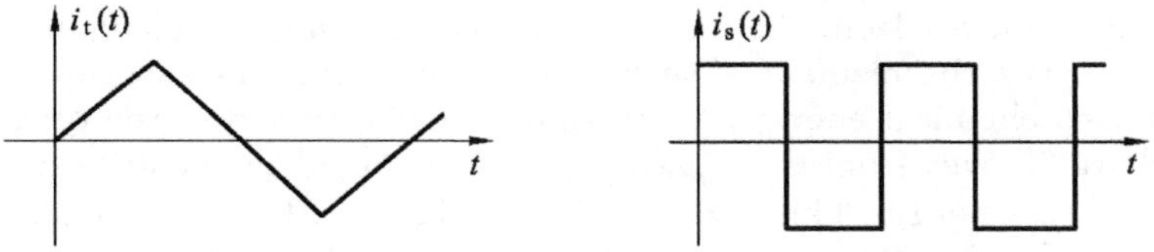

(a) Triangular waveform (b) Square waveform

Figure 1.1.4 AC current can have various waveforms

1.1.2.4 Double-subscript nation for currents

So far we have used arrows along circuit elements or conductors to indicate reference directions for currents. Another way to indicate the current and reference direction for a circuit element is to label the ends of the element and use double subscripts to define the reference direction for the current. For example, consider the resistance of Figure 1.1.5.

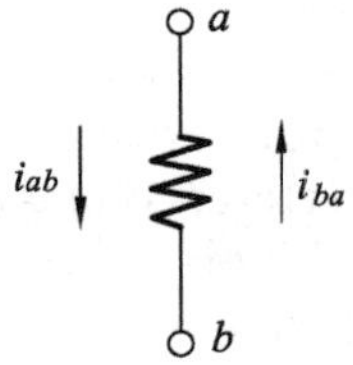

Figure 1.1.5 Reference directions can be indicated by labeling the ends of circuit elements and using double subscripts on current variables

The current denoted by i_{ab} is the current through the element with its reference direction pointing from a to b. Similarly, i_{ba} is the current with its reference direction from b to a. Of course, i_{ab} and i_{ba} are the same in magnitude and opposite in sign, because they denote the same current but with opposite reference directions. Thus we have $i_{ab}=-i_{ba}$.

1.1.3 Voltage

When charge moves through circuit elements, energy can be transferred. In the case of automobile headlights, stored chemical energy is supplied by the battery and absorbed by the headlights where it appears as heat and light. The voltage associated with a circuit element is the energy transferred per unit of charge that flows through the element. The units of voltage are volts (V), which are equivalent to joules per coulomb (J/C). We express this ratio in differential form as

$$V=\frac{\mathrm{d}w}{\mathrm{d}q} \tag{1.1.2}$$

For example, consider the storage battery in an automobile. The voltage across its terminals is (nominally) 12 V. This means that 12 J are transferred to or from the battery for each coulomb that flows through it. When charge flows in one direction, energy is supplied by the battery, appearing elsewhere in the circuit as heat or light or perhaps as mechanical energy at the starter motor. If charge moves through the battery in the opposite direction, energy is absorbed by the battery, where it appears as stored chemical energy.

Voltages are assigned polarities that indicate the direction of energy flow. If positive charge moves from the positive polarity through the element toward the negative polarity, the element absorbs energy that appears as heat, mechanical energy, stored chemical energy, or as some other forms. On the other hand, if positive charge moves from the negative polarity toward the positive polarity, the element supplies energy. This is illustrated in Figure 1.1.6. For negative charge, the direction of energy transfer is reversed.

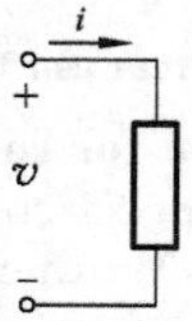

Figure 1.1.6 When current flows through an element and voltage appears across the element, energy is transferred

1.1.3.1 Reference polarities

When we begin to analyze a circuit, we often do not know the actual polarities of some of the voltages of interest in the circuit. Therefore, we simply assign voltage variables choosing reference polarities arbitrarily. (Of course, the actual polarities are not arbitrary). This is illustrated in Figure 1.1.5. Next, we apply circuit principles, obtaining equations that are solved for the voltages. If a given voltage has an actual polarity opposite to our arbitrary choice for the reference polarity, we obtain a negative value for the voltage. For example, if we find that $v_3=-5$ V in Figure 1.1.2, we know that the voltage across element 3 is 5 V in magnitude and its actual polarity is opposite to that shown in the Figure (i.e., the actual polarity is positive at the bottom end of element 3 and negative at the top).

Usually we do not put much effort into trying to assign "correct" references for current directions or voltage polarities to determine true directions and polarities (as well as magnitudes of the currents and voltages).

Voltages can be constant with time or they can vary. Constant voltages are called DC voltages. On the other hand, voltages that change in magnitude and alternate in polarity with time are said to be AC voltages. For example,

$$V_1(t)=10\ \text{V}$$

It is a DC voltage. It has the same magnitude and polarity for all time. On the other hand,

$$V_2(t)=10\cos(200t)\text{V}$$

It is an AC voltage that varies in magnitude and polarity. When assuming $V_2(t)$ a negative value, the actual polarity is opposite the reference polarity.

1.1.3.2 Double-subscript nation for voltages

Another way to indicate the reference polarity of a voltage is to use double-subscripts on the voltage variable. We use letters or numbers to label the terminals between which the voltage appears, as illustrated in Figure 1.1.7.

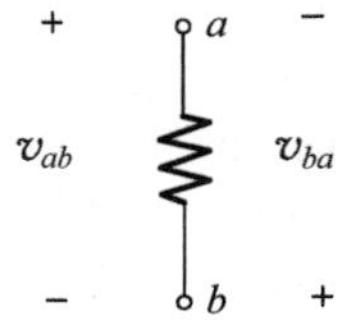

Figure 1.1.7 The voltage v_{ab} has a reference polarity that is positive at point a and negative at point b

For the resistance shown in Figure 1.1.7, v_{ab} represents the voltage between point a and b with the positive reference at point a. The two subscripts identify the points between which the voltage appears and the first subscript is the positive reference. Similarly, v_{ba} represents the voltage between point b and a with the positive reference at point b. Thus, we can write $v_{ab}=-v_{ba}$.

Still another way to indicate voltage and its reference polarity is to use an arrow. As shown in Figure 1.1.8, the positive reference corresponds to the head of the arrow.

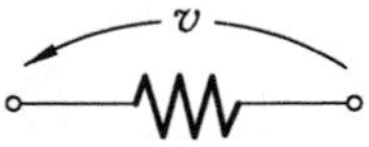

Figure 1.1.8 The positive reference for v is at the head of the arrow

Example 1.1.1

No charge exists at the upper terminal of the element in Figure for example 1.1.1 for $t<0$. At $t=0$, a 5 A current begins to flow into the upper terminal.

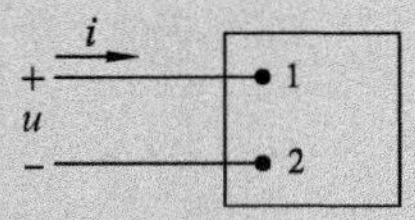

Figure for example 1.1.1

(a) Derive the expression for the charge accumulating at the upper terminal of the element for $t>0$.

(b) If the current is stopped after 10 seconds, how much charge has accumulated at the upper terminal?

Solution: (a) From the definition of current given in Equation (1. 1. 1), the expression for charge accumulation due to current flow is $q(t) = \int_0^t i(x)\mathrm{d}x$.

Therefore, $q(t) = \int_0^t 5\mathrm{d}x = 5t - 5(0) = 5t$ C for $t>0$.

(b) The total charge that accumulates at the upper terminal in 10 seconds due to a 5 A current is

$$q(t)=5(10)=50 \text{ C}$$

1.2 Power and energy

Power and energy calculations are also important in circuit analysis. One reason is that although voltage and current are useful variables in the analysis and design of electrically based systems, the useful output of the system often is nonelectrical. But this output is conveniently expressed in terms of power or energy. Another reason is that all practical devices have limitations on the amount of power that they can handle. In the design process, therefore, voltage and current calculations by themselves are not sufficient.

We now relate power and energy to voltage and current, and at the same time use the power calculation to illustrate the passive sign convention. Recall from basic physics that power is the time rate of expending or absorbing energy. Mathematically, energy per unit time is expressed in the form of a derivative, or

$$P=\frac{\mathrm{d}w}{\mathrm{d}t} \tag{1. 2. 1}$$

The power associated with the flow of charge follows directly from the definition of voltage and current in Equations(1. 1. 1) and (1. 1. 2), or

$$P=\frac{\mathrm{d}w}{\mathrm{d}t}=\left(\frac{\mathrm{d}w}{\mathrm{d}q}\right)\left(\frac{\mathrm{d}q}{\mathrm{d}t}\right)$$

So

$$p=vi \tag{1. 2. 2}$$

Equation (1. 2. 2) shows that the power associated with a basic circuit element is simply the product of the current in the element and the voltage across the element. Therefore, power is a quantity associated with a pair of terminals, and we have to be able to tell from our calculation whether power is being delivered to the pair of terminals or extracted from it. This information comes from the correct application and interpretation of the passive sign convention.

If we use the passive sign convention, Equation (1. 2. 2) is correct if the reference direction for the current is in the direction of reference voltage drop across the terminals. Otherwise, Equation (1. 2. 2) must be written with a minus sign. In other words, if the current reference is in direction of a reference voltage rise across the terminals, the expression for the

$$p=-vi \tag{1. 2. 3}$$

The algebraic sign of power is based on charge movement through voltage drops and rises. As positive charges move through a drop in voltage, they lose energy, and as they move through a rise in voltage, they gain energy. Figure 1. 2. 1 summarizes the relationship between the polarity references for voltage and current and the expression for power.

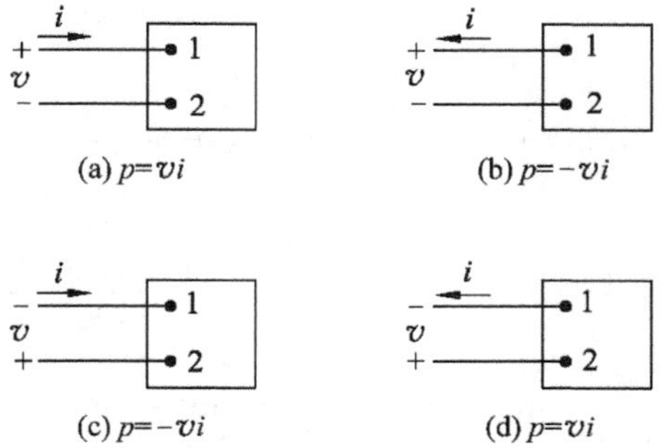

Figure 1. 2. 1 Polarity references and the expression for power

If the power is positive (that is, if $p>0$), power is being delivered to the circuit inside the box. If the power is negative (that is, if $p<0$), power is being extracted from the circuit inside the box.

For example, suppose that we have selected the polarity references shown in Figure 1. 2. 1 (b). Assume further that our calculations for the current and voltage yield the following numerical results:

$$i=4\text{ A and } v=-10\text{ V}$$

Then the power associated with the terminal pair 1,2 is

$$p=-(-10)(4)=40\text{ W}$$

Thus, the circuit inside the box is absorbing 40 W.

To take this analysis one step further, assume that a colleague is solving the same problem but has chosen the reference polarities shown in Figure 1. 2. 1(c). The numerical values are

$$i=-4\text{ A}, v=10\text{ V and } p=-(10)(-4)=40\text{ W}$$

Note that interpreting these results in terms of the reference system gives the same conclusions that we preciously obtained—namely, that the circuit inside the box is absorbing 40 W. In fact, any of the reference systems in Figure 1. 2. 1 yields the same result.

Example 1.2.1

Assume that the voltage at the terminals of the element in Figure 1.2.1(a)

$$i=20e^{-5\,000t}\text{ A},\ v=10e^{-5\,000t}\text{ kV}$$

(a) Calculate the power supplied to the element at 1 ms.

(b) Calculate the total energy (in joules) delivered to the circuit element.

Solution (a) Since the current is entering the "+" terminal of the voltage drop defined for the element in Figure 1. 2. 1 (a), we use a "+" sign in the power equation.

$$p=vi=(10\,000e^{-5\,000t})(20e^{-5\,000t})=200\,000e^{-10\,000t}\text{ W}$$

$$p(0.001)=200\,000e^{-10\,000(0.001)}=200\,000e^{-10}=0.908\text{ W}$$

(b) From the definition of power given in Equation (1. 2. 2), the expression for energy is

$$w(t) = \int_0^t p(x)\mathrm{d}x$$

To find the total energy delivered, integrate the expression for power from zero to infinity.

Therefore,

$$\begin{aligned} w_{total} &= \int_0^{\infty} 200\ 000\mathrm{e}^{-10\ 000x}\mathrm{d}x = \left.\frac{200\ 000\mathrm{e}^{-1\ 000t}}{-10\ 000}\right|_0^{\infty} \\ &= -20\mathrm{e}^{-\infty} - (-20\mathrm{e}^{-0}) = 20\ \mathrm{J} \end{aligned}$$

Thus, the total energy supplied to the circuit element is 20 J.

1.3 Voltage and current sources

1.3.1 Electrical source

An electrical source is a device that is capable of converting nonelectric energy to electric energy and vice versa. A discharging battery converts chemical energy to electric energy, whereas a battery being charged converts electric energy to chemical energy. A dynamo is a machine that converts mechanical energy to electric energy and vice versa. If operating in the mechanical to electric mode, it is called a generator. If transforming from electric to mechanical energy, it is referred to as a motor. The important thing to remember about these sources is that they can either deliver or absorb electric power, generally maintaining either voltage or current. This behavior is of particular interest for circuit analysis and led to the creation of the ideal voltage source and ideal current source as basic circuit elements.

1.3.2 Ideal independent sources

An ideal voltage source is a circuit element that maintains a prescribed voltage across its terminals regardless of the current flowing in those terminals. Similarly, an ideal current source is a circuit element that maintains a prescribed current through its terminals regardless of the voltage across those terminals. These elements do not exist as practical devices—they are idealized models of actual voltage and current sources.

Using an ideal model for current and voltage sources places an important restriction on how we may describe them mathematically. Because an ideal voltage source provides a steady voltage, even if the current in the element changes, it is impossible to specify the current in an ideal voltage source as a function of its voltage. Likewise, if the only information you have about an ideal current source is the value of current supplied, it is impossible to determine the voltage across that current source. We have sacrificed our ability to relate voltage and current in a practical source for the simplicity of using ideal sources in circuit analysis.

Ideal voltage and current sources can be further described as either independent sources or dependent sources. An independent source establishes a voltage or current in a circuit without relying on voltages or currents elsewhere in the circuit. The value of the voltage or current supplied is specified by the value of the independent source alone. In contrast, a dependent source establishes a voltage or current whose value depends on the value of a voltage or current elsewhere in the circuit. You cannot specify the value of a dependent source unless you know the value of the voltage or current on which it depends.

The circuit symbols for the ideal independent sources are shown in Figure 1. 3. 1. Note that a circle is used to represent an independent source in a circuit, and you must include the value of the supplied voltage and reference polarity, as shown in Figure 1. 3. 1 (a). Similarly, to completely specify an ideal independent current source, you must include the value of the supplied current and its reference direction, as shown in Figure 1. 3. 1(b).

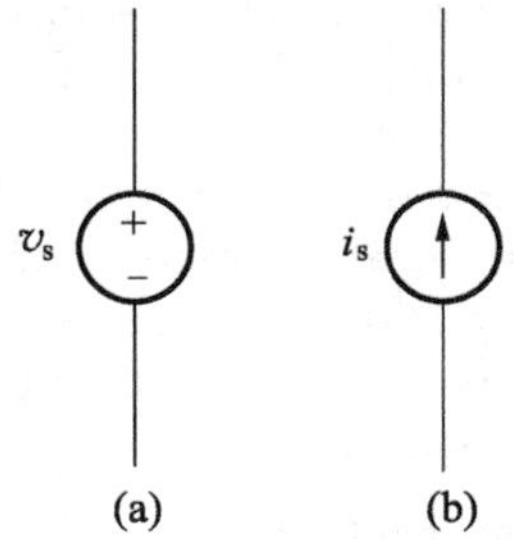

Figure 1. 3. 1 The circuit symbols for (a) an ideal independent voltage source and (b) an ideal independent current source

Example 1.3.1

Using the definition of the ideal independent voltage and current sources, state which interconnections in Figure for example 1. 3. 1 are permissible and which violate the constraints imposed by the ideal sources.

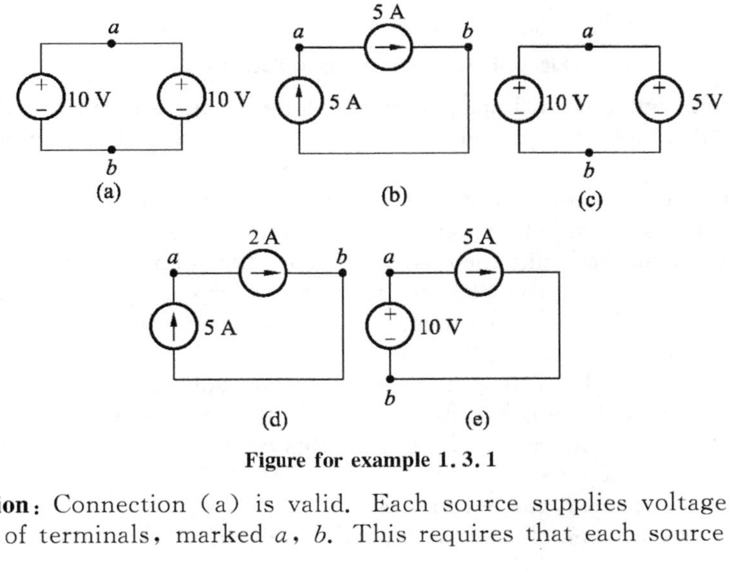

Figure for example 1. 3. 1

Solution: Connection (a) is valid. Each source supplies voltage across the same pair of terminals, marked a, b. This requires that each source supply the same voltage with the same polarity, which they do.

Connection (b) is valid. Each source supplies current through the same pair of terminals, marked a, b. This requires that each source supply the same current in the same direction, which they do.

Connection (c) is not permissible. Each source supplies voltage across the same pair of terminals, marked a, b. This requires that each source supply the same voltage with the same polarity, which they do not.

Connection (d) is not permissible. Each source supplies current through the same pair of terminals, marked a, b. This requires that each source supply the same current in the same direction, which they do not.

Connection (e) is valid. Voltage source supplies voltage across the pair of terminals, marked a, b. The current source supplies current through the same pair of terminals. Because an ideal voltage source supplies the same voltage regardless of the current, and an ideal current source supplies the same current regardless of the voltage, this is a permissible connection.

1.3.3 Ideal dependent sources

1.3.3.1 Dependent voltage sources

A dependent or controlled voltage source is similar to an independent source except that the voltage across the source terminal is a function of other voltages or currents in the circuit. Instead of a circle, it is customary to use a diamond-shaped box to represent controlled sources in circuit diagrams. Two examples of dependent voltage sources are shown in Figure 1.3.2.

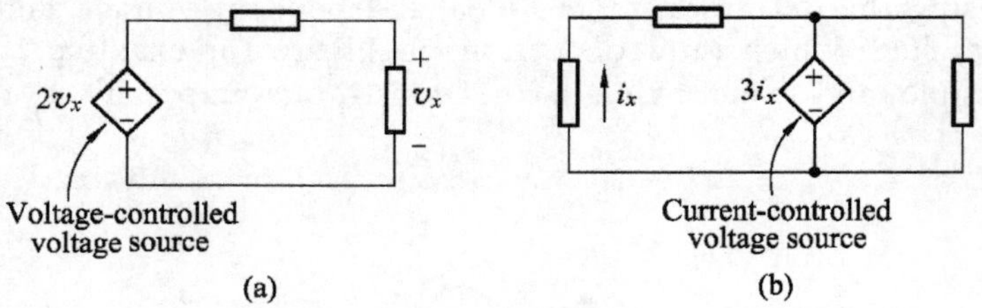

Figure 1.3.2 Dependent voltage sources

A voltage-controlled voltage source is a voltage source having a voltage equal to a constant times the voltage across a pair of terminals elsewhere in the network.

An example is shown in Figure 1.3.2 (a). The dependent voltage source is represented by a diamond symbol. The reference polarity of the source is indicated by the marks inside the diamond-shaped box. The voltage v_x determines the value of the voltage produced by the source. For example, if it should turn out that $v_x=2$ V, the source voltage is $2v_x=4$ V. If v_x should equal -3 V, the source produces $2v_x=-6$ V.

A current-controlled voltage source is a voltage source having a voltage equal to a constant times the current through some other elements in the circuit. An example is shown in Figure 1.3.2 (b). In this case, the source voltage is three times the value of the current i_x.

1.3.3.2 Dependent current sources

The current flowing through a dependent current source is determined by a current or voltage elsewhere in the circuit. The symbol is a diamond-shaped box enclosing an arrow that indicates the reference direction. Two types of controlled current sources are shown in Figure 1. 3. 3.

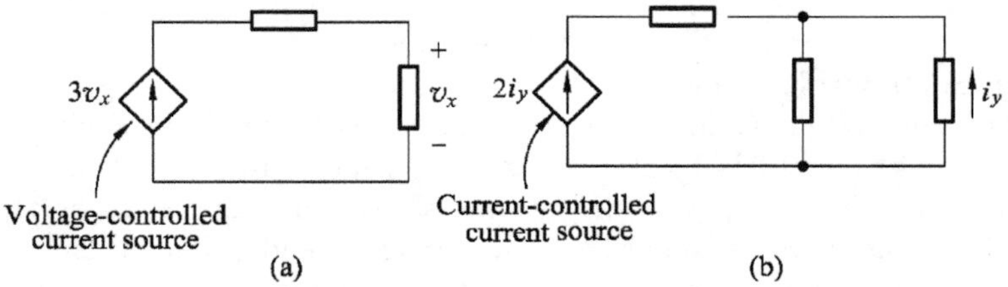

Figure 1. 3. 3 Dependent current sources

In Figure 1. 3. 3 (a), we have a voltage-controlled current source. The current through the source is three times the voltage v_x. Figure 1. 3. 3 (b) illustrates a current-controlled current source. In this case, the current through the source is twice the value of i_y.

Like controlled voltage source, controlled current sources are useful in constructing circuit models for many types of real-word devices, such as electronic amplifiers, transistors, transformers, and electrical machines. If a controlled source is needed for some application, it can be implemented by using electronic amplifiers. In sum, these are the four kinds of controlled sources.

Example 1.3.2

Using the definitions of the ideal independent and dependent sources, state which interconnections in Figure for example 1. 3. 2 are valid and which violate the constraints imposed by the ideal sources.

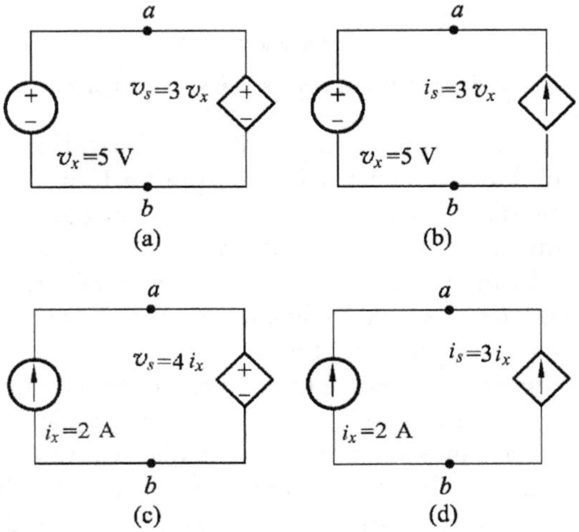

Figure for example 1. 3. 2

Solution: Connection (a) is invalid. Both the independent source and the dependent source supply voltage across the same pair of terminals, labeled a, b. This requires that each source supply the same voltage with the same polarity. But

$$v_s = 3v_x = 15 \text{ V}$$

this is not an allowable connection.

Connection (b) is valid. The independent voltage source supplies voltage across the pair of terminals marked a, b. The dependent current source supplies current through the same pair of terminals. Because an ideal voltage source supplies the same voltage regardless of current, and an ideal current source supplies the same current regardless of voltage, this is an allowable connection.

Connection (c) is valid. The independent current source supplies current through the pair of terminals marked a, b. The dependent voltage source supplies voltage across the same pair of terminals. Because an ideal current source supplies the same current regardless of voltage, and an ideal voltage source supplies the same voltage regardless of current, this is an allowable connection.

Connection (d) is invalid. Both the independent source and the dependent source supply current through the same pair of terminals, labeled a, b. This requires that each source supply the same current in the same reference direction. But

$$i_s = 3i_x = 6 \text{ A}$$

this is not an allowable connection.

1.4 Electrical resistance (Ohm's law)

1.4.1 Resistance and Ohm's law

Resistance is the capacity of materials to impede the flow of current or more specifically, the flow of electric charge. The circuit element used to mode this behavior is the resistor. Figure 1.4.1 shows the circuit symbol for a resistor, with R denoting the resistance value of a resistor.

Figure 1.4.1 The circuit symbol for a resistor having a resistance *R*

Conceptually, we can understand resistance if we think about the moving electrons that make up electric current interacting with and being resisted by the atomic structure of the material through which they are moving. In the course of these interactions, some amount of electric energy is converted to thermal energy and dissipated in the form of heat. This effect may be undesirable. However, many useful electrical devices take advantage of resistance heating, including stoves, toasters, irons, and space heaters.

Most materials exhibit measurable resistance to current. The amount of resistance depends on the material. Metals such as copper and aluminum have small values of resistance, making them good choices for wiring used to conduct electric current. In fact, when represented in a circuit diagram, copper or aluminum wiring isn't usually modeled as a resistor; the resistance of the wire is so small compared with the resistance of other elements in the circuit that we can neglect the wiring resistance to simplify the diagram.

For purposes of circuit analysis, we must reference the current in the resistor to the terminal voltage. We can do so in two ways: either in the direction of the voltage drop across the resistor, or in the direction of the voltage rise across the resistor, as shown in Figure 1. 4. 2. and Figure 1. 4. 3.

(a) Resistance symbol

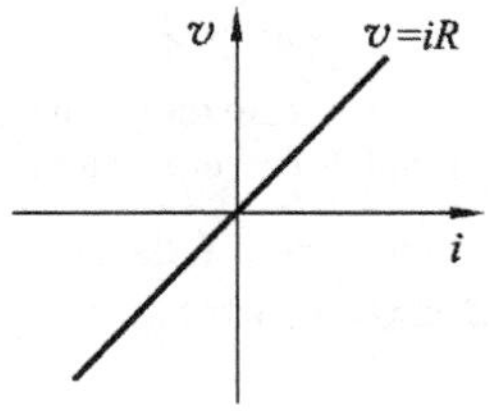

(b) Ohm's law

Figure 1. 4. 2 Voltage is proportional to current in an ideal resistor

If we choose Figure 1. 4. 2, the relationship between the voltage and current is related by Ohm's law:

$$v=iR \tag{1.4.1}$$

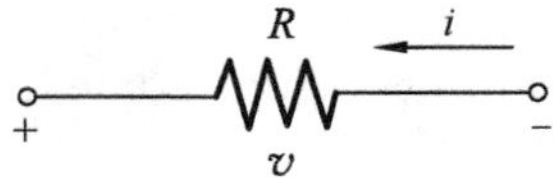

Figure 1. 4. 3 The references for *v* and *i* are opposite to the passive configuration

If we choose the Figure 1. 4. 3, current reference direction enters the negative reference of the voltage, and Ohm's law becomes:

$$v=-iR \tag{1.4.2}$$

1.4.2 Conductance

Ohm's law expresses the voltage as a function of the current. However, expressing the current as a function of the voltage is also convenient. Thus,

$$i=\frac{v}{R}$$

The reciprocal of the resistance is referred to as conductance. It is customary to denote conductance with the letter G:

$$G=\frac{1}{R} \tag{1.4.3}$$

Conductance has the units of inverse ohms (Ω^{-1}), which are called siemens. Thus, we can write Ohm's law as

$$i=Gv \tag{1.4.4}$$

1.4.3 Resistor

The dimensions and geometry of the resistor as well as the particular material used to construct a resistor influence its resistance. We consider only resistors that take the form of a long cylinder or bar with terminals attached at the ends, as illustrated in Figure 1. 4. 4.

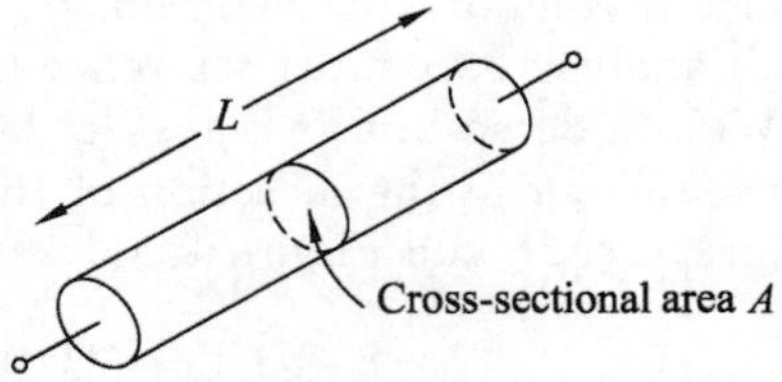

Figure 1.4.4 Resistors often take the form of a long cylinder (or bar) in which current enters one end and flows along the length

The cross-section area A is constant along the length of the cylinder section, the resistance is approximately given by

$$R=\rho\frac{L}{A} \tag{1.4.5}$$

We use ideal resistors in circuit analysis to model the behavior of physical devices. Using the qualifier ideal model reminds us that the resistor model makes several simplifying assumptions about the behavior of actual resistive devices. The most important of these simplifying assumptions is that the resistance of the ideal resistor is constant and its value does not vary over time. Most actual resistive devices do not have constant resistance, and their resistance does vary over time. The ideal resistor model can be used to represent a physical device whose resistance doesn't vary much from some constant value over the time period of interest in the circuit analysis.

1.4.4 Power calculations for resistances

We may calculate the power at the terminals of a resistor in several ways. The first approach is to use the defining equation and simply calculate the product of the terminal voltage and current. We write

$$p=vi \tag{1.4.6}$$

when $v=iR$ and

$$p=-vi \tag{1.4.7}$$

when $v=-iR$

A second method of expressing the power at the terminals of a resistor is in terms of the current and the resistance.

$$p=i^2R \tag{1.4.8}$$

A third method of expressing the power at the terminals of a resistor is in terms of the voltage and resistance. Expression is independent of the polarity references, so

$$p=\frac{v^2}{R} \tag{1.4.9}$$

Sometimes a resistor's value will be expressed as a conductance rather than as a resistance. Using the relationship between resistance and conductance given in Equation (1.4.3), we may also write power in terms of the conductance,

$$p=\frac{i^2}{G} \text{ or } p=v^2G$$

Notice that power for a resistance is positive regardless of the sign of v or i. Thus, power is absorbed by resistances. If the resistance results from collisions with the atoms of the material composing a resistor, this power shows up as heat.

Example 1.4.1

In each circuit in Figure for example 1.4.1, either the value of v or i is not known.

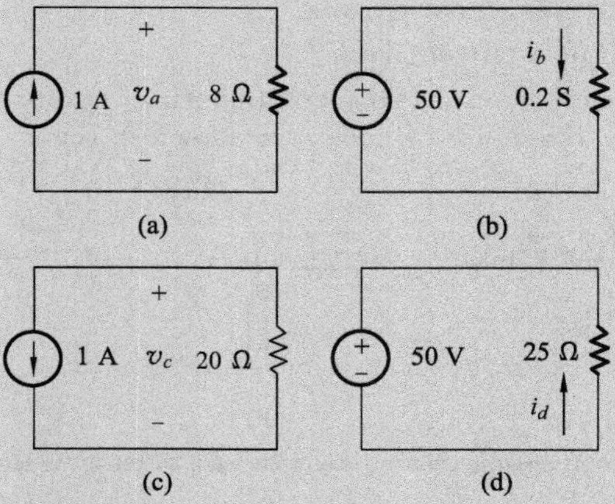

Figure for example 1.4.1

(a) Calculate the values of v and i.
(b) Determine the power dissipated in each resistor.

Solution: (a) The voltage v_a in Figure for example 1.4.1(a) is a drop in the direction of the current in the resistor. Therefore,

$$v_a=(1)(8)=8\ \text{V}$$

The current i_b in the resistor with a conductance of 0.2 S in Figure for example 1.4.1(b) is the direction of the voltage drop across the resistor. Thus,

$$i_b=(50)(0.2)=10\ \text{A}$$

The voltage v_c in Figure for example 1.4.1(c) is a rise in the direction of the current in the resistor. Hence

$$v_c=-(1)(20)=-20\ \text{V}$$

The current i_d in the 25 Ω resistor in Figure for example 1.4.1(d) is in the direction of the voltage rise across the resistor. Therefore,

$$i_d=\frac{-50}{25}=-2\ \text{A}$$

(b) The power dissipated in each of the four resistors is

$$p_{8\ \Omega}=\frac{(8)^2}{8}=8\ \text{W}$$

$$p_{0.2\ \text{s}}=(50)^2(0.2)=500\ \text{W}$$

$$p_{20\ \Omega}=\frac{(-20)^2}{20}=(1)^2(20)=20\ \text{W}$$

$$p_{25\ \Omega}=\frac{(50)^2}{25}=(-2)^2(25)=100\ \text{W}$$

1.5 Kirchhoff's law

A circuit is said to be solved when the voltage across and the current in every element have been determined. Ohm's law is an important equation for deriving such solution. However, Ohm's law may not be enough to provide a complete solution. We need to use more important algebraic relationships, known as Kirchhoff's laws, to solve most circuits.

1.5.1 Kirchhoff's current law

A note in an electrical circuit is a point at which two or more circuit elements are joined together. Examples of nodes are shown in Figure 1. 5. 1.

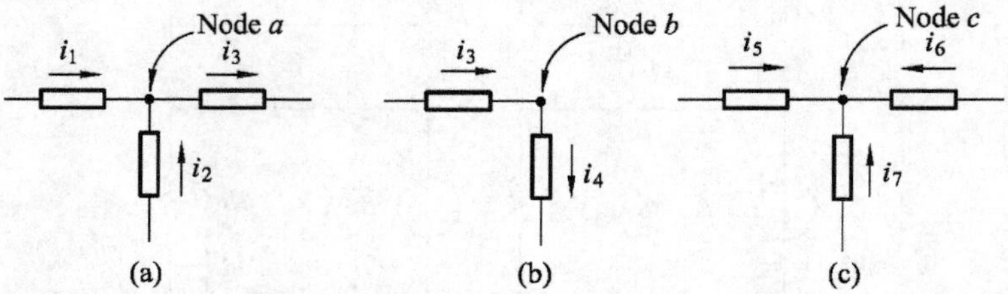

Figure 1. 5. 1 Partial circuits showing one node each to illustrate Kirchhoff's current law

An important principle of electrical circuit is Kirchhoff's current law: the net current entering a node is zero. To compute the net current entering a node, we add the currents entering and subtract the currents leaving. For illustration, consider the nodes of Figure 1. 5. 1. Then, we can write:

$$\text{node } a: i_1 + i_2 - i_3 = 0$$
$$\text{node } b: i_3 - i_4 = 0$$
$$\text{node } c: i_5 + i_6 + i_7 = 0$$

Notice that for node b, Kirchhoff's current law requires that $i_3 = i_4$. In general, if only two circuit elements are connected at a node, their currents must be equal. The current flows into the node through one element and out through the other. Usually, we will recognize this fact and assign a single current variable for both circuit elements.

We abbreviate Kirchhoff's current law as KCL. There is another way to state KCL: the sum of the current entering a node equals the sum of the currents leaving a node. Applying this statement to Figure 1. 5. 1, we obtain the following set of equations:

$$\text{node } a: i_1 + i_2 = i_3$$
$$\text{node } b: i_3 = i_4$$
$$\text{node } c: i_5 + i_6 + i_7 = 0$$

To use KCL, an algebraic sign corresponding to a reference direction must be assigned to every current at the node. Assigning a positive sign to a current leaving a node requires assigning a negative sign to a current entering a node. Conversely, giving a negative sign to a current leaving a node requires giving a positive sign to a current entering a node.

1.5.2 Kirchhoff's voltage law

Before we can state Kirchhoff's voltage law (KVL), we must define a loop. A loop in an electrical circuit is a closed path starting at a node and proceeding through circuit elements, eventually returning to the starting node. Frequently, several loops can be identified for a given circuit. For example, in Figure 1.5.2, one loop consists of the path starting at the top end of element A and proceed clockwise through elements B and C, returning through A to the starting point. Another loop starts at the top of element D and proceeds clockwise through E, F, and G, returning to the start through D. Still another loop exists through elements A, B, E, F, and G around the periphery of the circuit.

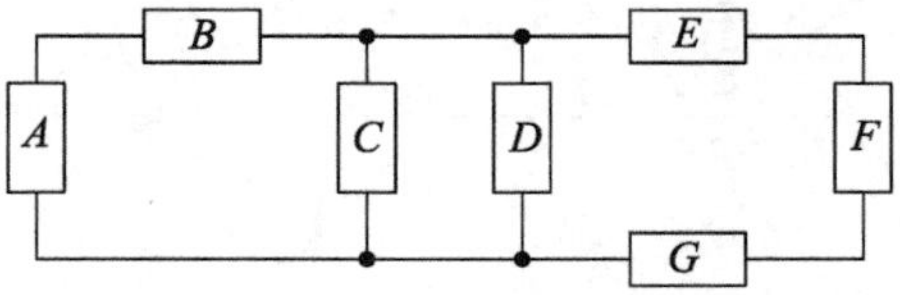

Figure 1.5.2 Circuit has many loops

KVL states: the algebraic sum of the voltage equals zero for any closed path (loop) in an electrical circuit. In traveling around a loop, we encounter various voltages, some of which carry a positive sign while others carry a negative sign in the algebraic sum. A convenient convention is to use the first polarity mark encountered for each voltage to decide if it should be added or subtracted in the algebraic sum. If we go through the voltage from the positive polarity reference in the opposite direction (minus to plus), the voltage carries a negative sign. This is illustrated in Figure 1.5.3.

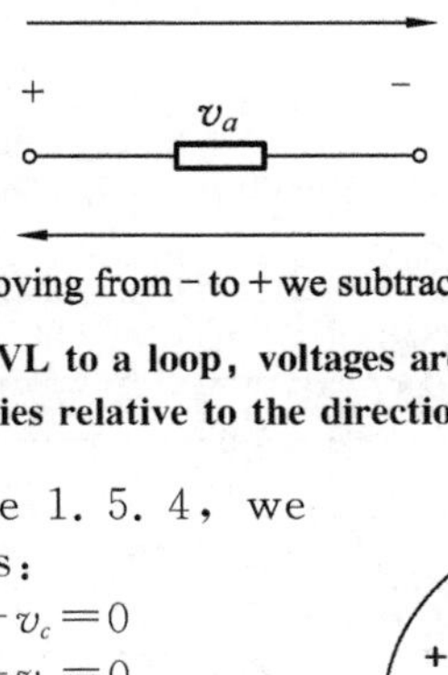

Figure 1.5.3 In applying KVL to a loop, voltages are added or subtracted depending on their reference polarities relative to the direction of travel around the loop

For the circuit of Figure 1.5.4, we obtain the following equations:

Loop 1: $-v_a+v_b+v_c=0$

Loop 2: $-v_c-v_d+v_e=0$

Loop 3: $v_a-v_b+v_d-v_e=0$

Notice that v_a is subtracted for Loop 1, but it is added for Loop 3, because the direction of travel is different for the two loops. Similarly, V_c is added for Loop 1 and subtracted for Loop 2.

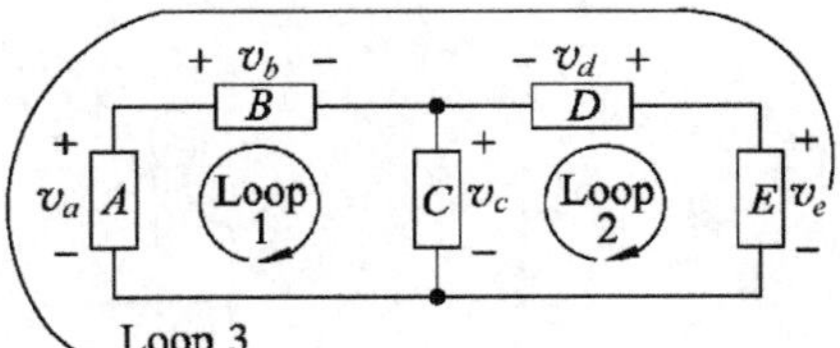

Figure 1.5.4 Circuit used for illustration of KVL

Example 1.5.1

Sum up the currents at each node in the circuit shown in Figure for example 1.5.1.

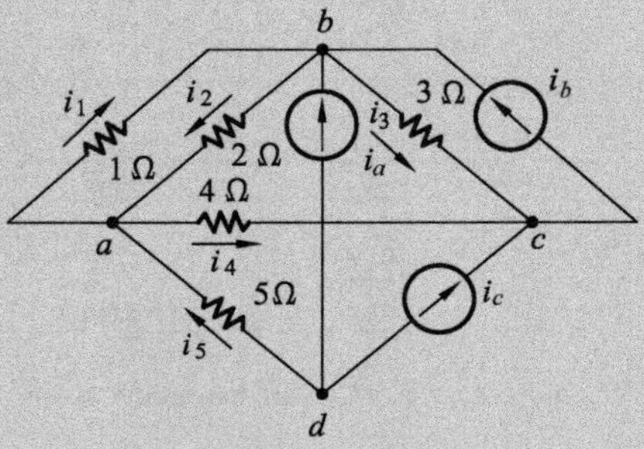

Figure for example 1.5.1

Solution: In writing the equations, we use a positive sign for a current leaving *a* node. The four equations are

node a: $i_1+i_4-i_2-i_5=0$

node b: $i_2+i_3-i_1-i_b-i_a=0$

node c: $i_b-i_3-i_4-i_c=0$

node d: $i_5+i_a+i_c=0$

Example 1.5.2

Sum up the voltages around each designed path in the circuit shown in Figure for example 1.5.2.

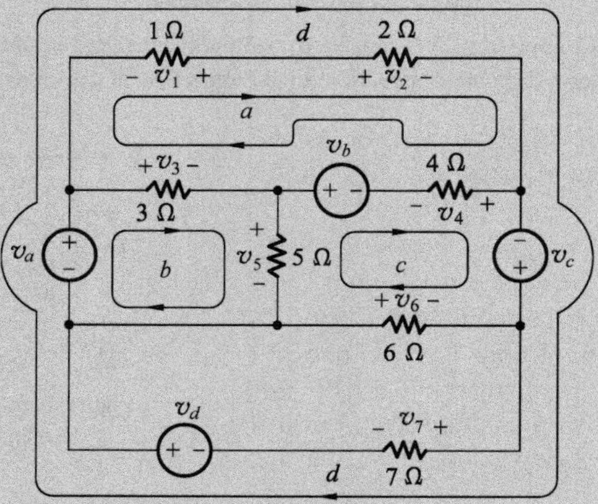

Figure for example 1.5.2

Solution: In writing the equations, we use a positive sign for a voltage drop. The four equations are

$$\text{loop } a: \ -v_1+v_2+v_4-v_b-v_3=0$$
$$\text{loop } b: \ -v_a+v_3+v_5=0$$
$$\text{loop } c: \ v_b-v_4-v_c-v_6-v_5=0$$
$$\text{loop } d: \ -v_a-v_1+v_2-v_c+v_7-v_d=0$$

Example 1.5.3

(a) Use Kirchhoff's laws and Ohm's law to find the voltage as shown in Figure for example 1.5.3.

(b) Show that your solution is consistent with the constraint that the total power developed in the circuit equals the total power dissipated.

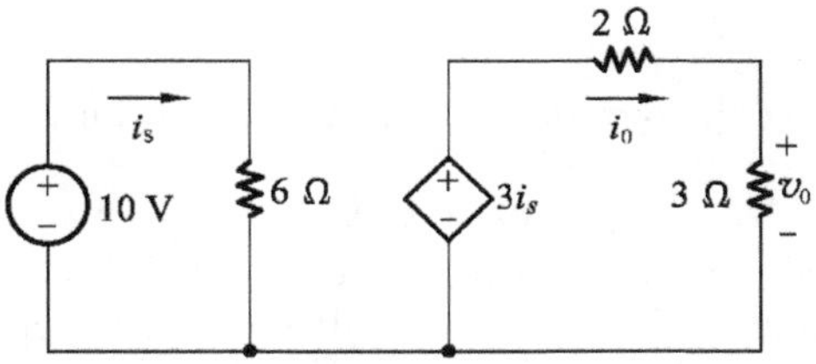

Figure for example 1.5.3

Solution: (a) There are two equations for the two currents. Apply the Kirchhoff's voltage law to the two closed loops, we can get

$$10=6i_s$$
$$3i_s=2i_0+3i_0$$

Solving for the current yields

$$i_s=1.67 \text{ A}$$
$$i_0=1 \text{ A}$$

Apply Ohm's law to the 3 Ω resistor gives the desired voltage:

$$v_0=3i_0=3 \text{ V}$$

(b) The power delivered to the independent voltage source is

$$p=-(10)(i_s)=-(10)(1.67)=-16.7 \text{ W}$$

The power delivered to the dependent voltage source is

$$p=(3i_s)(-i_0)=(5)(-1)=-5 \text{ W}$$

Both sources are developing power, and the total developed power is 21.7 W.

The power delivered to the 6 Ω resistor is

$$p=(6)(i_s^2)=(6)(1.67)^2=16.7 \text{ W}$$

The power delivered to the 2 Ω resistor is

$$p=(2)(i_0^2)=(2)(1)^2=2 \text{ W}$$

The power delivered to the 6 Ω resistor is

$$p=(3)(i_0^2)=(3)(1)^2=3 \text{ W}$$

The resistors all dissipate power, and the total power dissipated is 21.7 W, equal to the total power developed in the sources.

Summary

The objective of this chapter is to introduce the background needed in the following chapter for the analysis of linear resistive circuits.

1. Identify the principal elements of electric circuits: nodes, loops, voltage and current sources. These elements will be common to all electric circuits analyzed in the book.

2. Apply Ohm's and Kirchhoff's laws to simple electric circuits and derive the basic circuit equations. Mastery of these laws is essential to writing the correct equations for electric circuits.

3. Apply the passive sign convention and compute the power dissipated by circuit elements. The passive sign convention is a fundamental skill needed to derive the correct equations for an electric circuit.

Problems

1.1 The current at the terminals of the element in Figure for example 1.1.1 is

$$i=0 \qquad t<0$$
$$i=20e^{-5\,000t}\text{ A} \qquad t\geqslant 0$$

Calculate the total charge entering the element at its upper terminal.

1.2 Assume that a 20 V voltage drop occurs across an element from terminal 2 to terminal 1 and that a current of 4 A enters terminal 2.

(a) Specify the values of v and I for the polarity reference shown in Figure 1.2.1.

(b) State whether the circuit inside the box is absorbing or delivering power.

(c) How much power is the circuit absorbing?

1.3 The voltage and current at the terminals for the circuit element in Figure for example 1.1.1 are zero for $t<0$. For $t\geqslant 0$, they are

$$v=80\,000te^{-500t}\text{ V}$$
$$i=15te^{-500t}\text{ A}$$

(a) Find the time when the power delivered to the circuit element is maximum.

(b) Find the maximum value of power.

(c) Find the total energy delivered to the circuit element.

1.4 There is a high-voltage DC current transmission line between Shanghai and Nanjing, Nanjing is operating at 800 kV and carrying 1 800 A, as shown in Figure for problem 1.4. Calculate the power at the Shanghai end of the line and state the direction of power flow.

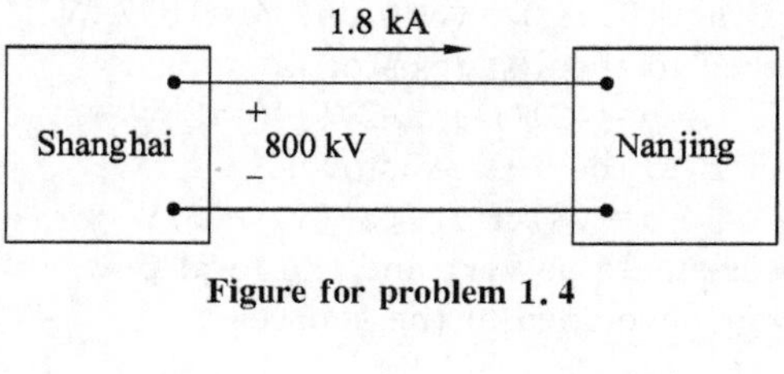

Figure for problem 1.4

1.5 For the circuit as shown in Figure for problem 1.5.

(a) What value of v_g is required in order for the interconnection to be valid?

(b) For this value of v_g, find the power associated with the 8 A source

1.6 For the circuit shown in Figure for problem 1.6.

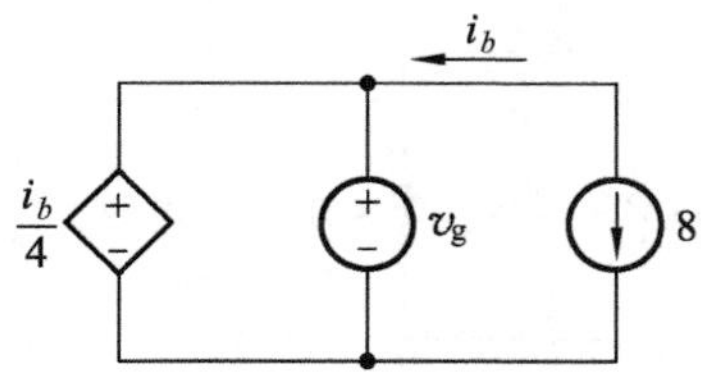

Figure for problem 1.5

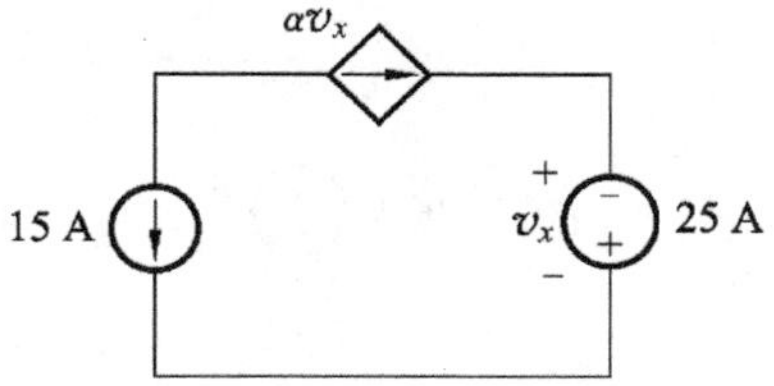

Figure for problem 1.6

(a) What value of α is required in order for the interconnection to be valid?

(b) For this value of α calculated in part (a), find the power associated with the 25 V source.

1.7 For the circuit shown in Figure for problem 1.7.

(a) If $v_g=1$ kV and $i_g=5$ mA, find the value of R and the power absorbed by the resistor.

(b) If $i_g=75$ mA and the power delivered by the voltage source is 3 W, find v_g, R, and the power absorbed by the resistor.

(c) If $R=300\ \Omega$ and the power absorbed by R is 480 mW, find i_g and v_g.

1.8 For the circuit shown in Figure for problem 1.8.

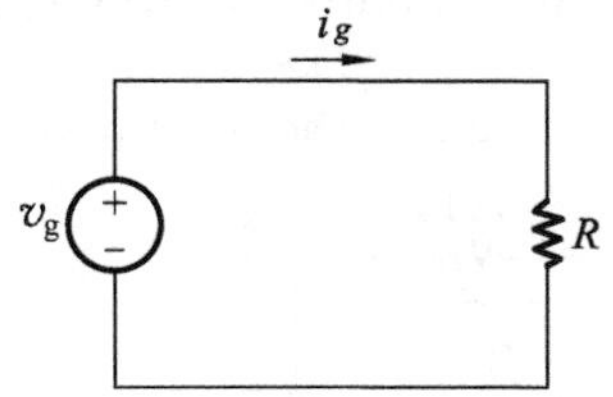

Figure for problem 1.7

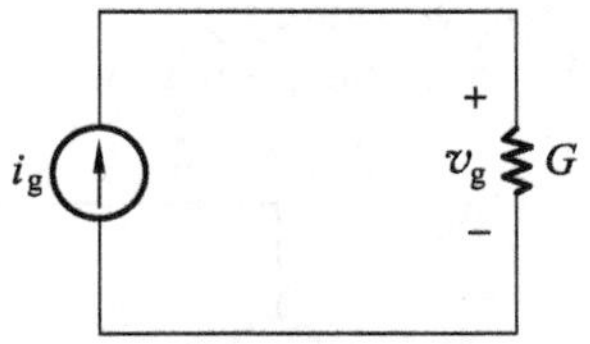

Figure for problem 1.8

(a) If $i_g=0.5$ A and $G=50$ mS, find v_g and power delivered by the current source.

(b) If $v_g=15$ V, and power delivered to the conductor is 9 W, find the conductance G and the source current i_g.

(c) If $G=200\ \mu$S and the power delivered to the conductance is 8 W, find i_g and v_g.

1.9 For the circuit shown in Figure for problem 1.9, calculate

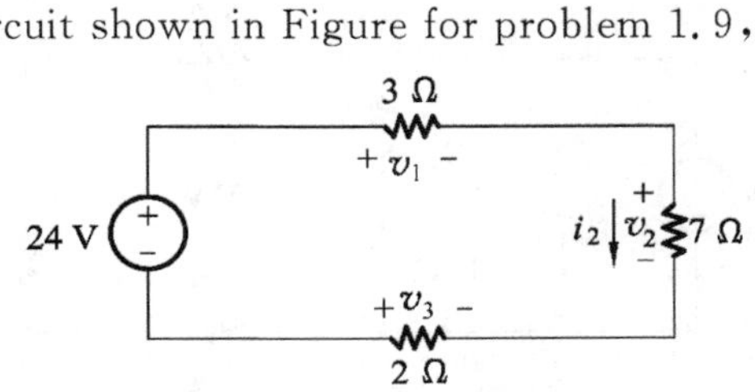

Figure for problem 1.9

(a) i_5, v_1, v_2, v_3.

(b) The power delivered by the 24 V source.

1.10 Use Ohm's law and Kirchhoff's laws to find the value of R in the circuit shown in Figure for problem 1.10.

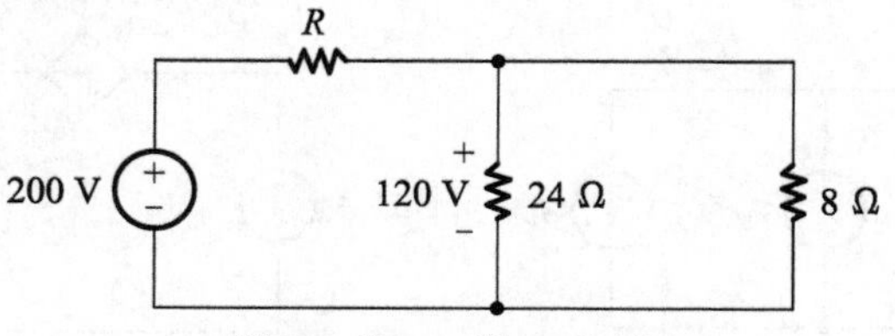

Figure for problem 1.10

1.11 For the circuit shown in Figure for problem 1.11.

(a) Find i.

(b) Find the power supplied by the voltage source.

(c) Reverse the polarity of the voltage source ad repeat parts (a) and (b).

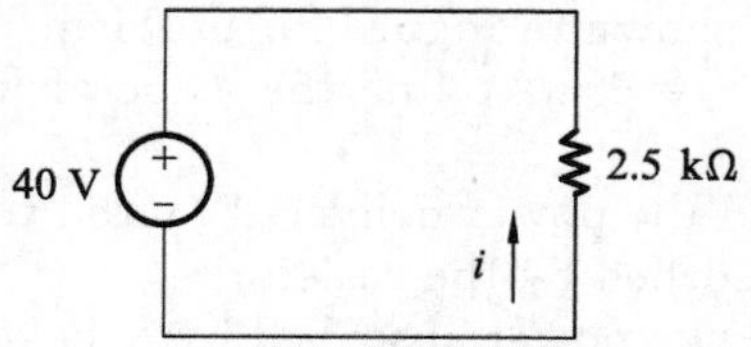

Figure for problem 1.11

1.12 (a) Find the currents in the circuit in Figure for problem 1.12.

(b) Find the voltage.

(c) Verify that the total power developed equals the total power dissipated.

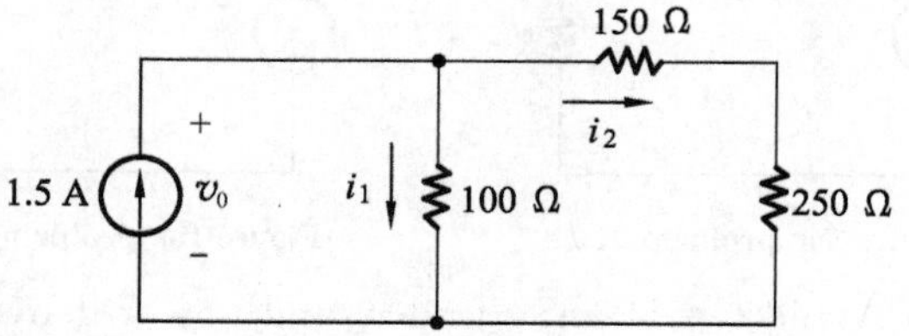

Figure for problem 1.12

1.13 For the circuit shown in Figure for problem 1.13, find

(a) R.

(b) The power supplied by the 240 V source.

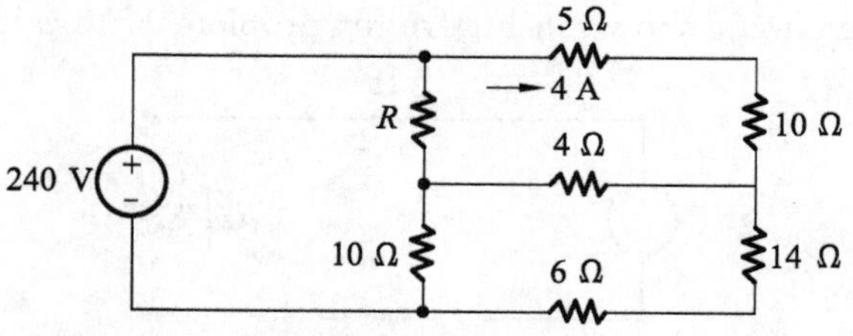

Figure for problem 1.13

1.14 The currents i_a and i_b in the circuit in Figure for problem 1.14 are 4 A and −2 A, respectively.

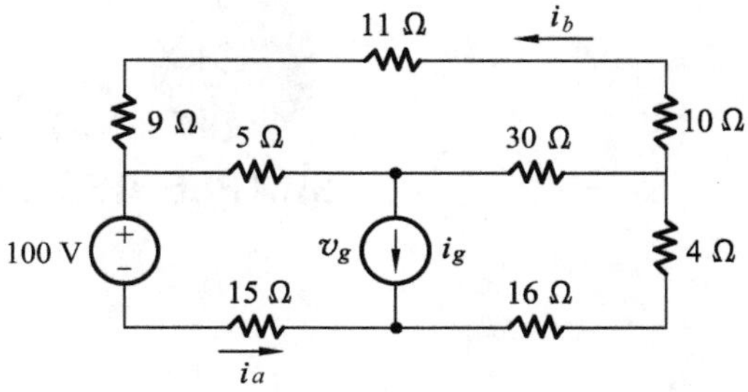

Figure for problem 1.14

(a) Find i_g.

(b) Find the power dissipated in each resistor.

(c) Find v_g.

(d) Show that the power delivered by the current source is equal to the power absorbed by all the other elements.

1.15 For the circuit shown in Figure for problem 1.15, find v_0 and the total power absorbed in the circuit.

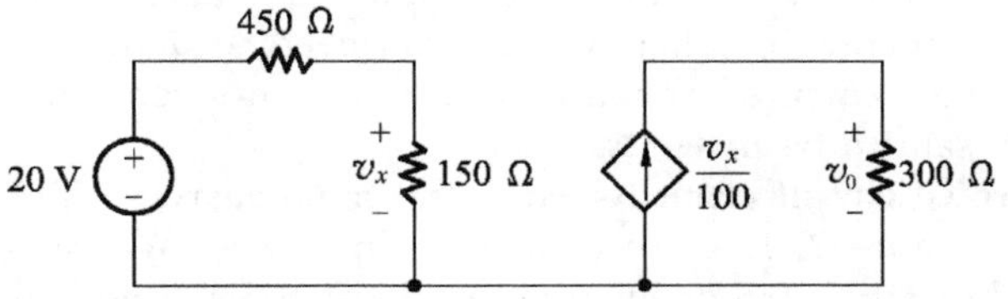

Figure for problem 1.15

CHAPTER 2
SIMPLE RESISTIVE CIRCUITS

Introduction

This chapter introduces some techniques commonly applied in circuit design and analysis. This techniques include combining resistors in series and parallel, voltage division and current division. The application of Ohm's law, Kirchhoff's law and the techniques will be restricted to resistive circuits in this chapter.

2.1 Resistors in series and parallel

2.1.1 Resistors in series

According to Kirchhoff's current law (KCL), when only two elements connect to a node, if you know the current in one of the elements, you also know it in the second element. In other words, you need to define only one unknown current for the two elements. When just two elements connect at a single node, the elements are said to be **in series.**

Series-connected circuit elements carry the same current. The resistors in the circuit shown in Figure 2.1.1 are connected in series. We can show that these resistors carry the same current by applying KCL to each node in the circuit.

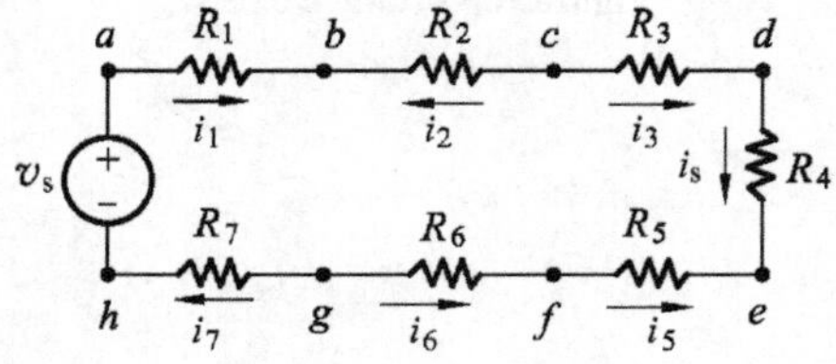

Figure 2.1.1 Resistors connected in series

$$i_s = i_1 = -i_2 = i_3 = i_4 = -i_5 = -i_6 = i_7 \qquad (2.1.1)$$

which states that if we know any one of the seven currents, we know them all. Thus we redraw Figure 2.1.1 as shown in Figure 2.1.2.

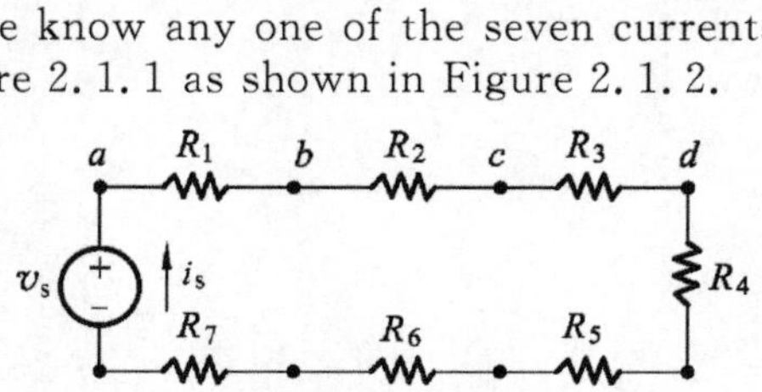

Figure 2.1.2 Series resistors with a single unknown current i_s

To find i_s, we apply KVL around the single closed loop. Defining the voltage across each resistor as a drop in the direction of i_s gives:

$$-v_s + i_s R_1 + i_s R_2 + i_s R_3 + i_s R_4 + i_s R_5 + i_s R_6 + i_s R_7 = 0$$

OR (2.1.2)

$$i_s (R_1 + R_2 + R_3 + R_4 + R_5 + R_6 + R_7) = v_s$$

We can use a single resistor to replace the seven resistors, whose numerical value is the sum of the individual resistors. That is

$$R_{eq} = R_1 + R_2 + R_3 + R_4 + R_5 + R_6 + R_7 \quad (2.1.3)$$

and

$$V_s = i_s R_{eq} \quad (2.1.4)$$

Thus we can redraw Figure 2.1.2 as shown in Figure 2.1.3

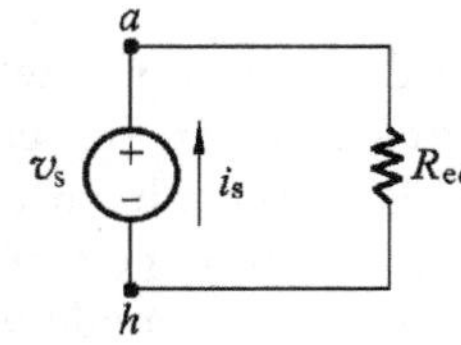

Figure 2.1.3 A simplified version of the circuit shown in Figure 2.1.2

In general, if k resistors are connected in series, the equivalent single resistor has a resistance equal to the sum of the k resistances, or

$$R_{eq} = \sum_{i=1}^{k} R_i = R_1 + R_2 + \cdots + R_k \quad (2.1.5)$$

Note that the resistance of the equivalent resistor is always larger than that of the largest resistor in the series connection.

Another way to think about this concept of an equivalent resistance is to visualize the string of resistors as being inside a black box. (An electrical engineer uses the term black box to imply an opaque container; that is the contents hidden from view. The engineer is then challenged to model the contents of the box by studying the relationship between the voltage and current at its terminals.) Determining whether the box contains k resistors or a single equivalent resistor is impossible. Figure 2.1.4 illustrates this method of studying the circuit shown in Figure 2.1.2.

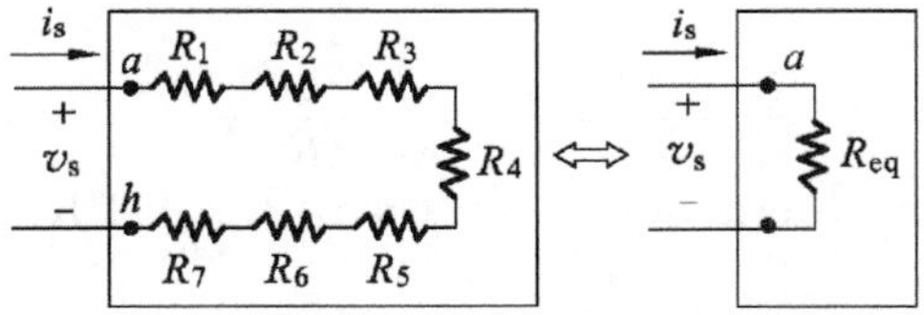

Figure 2.1.4 The black box equivalent of the circuit shown in Figure 2.1.2

2.1.2 Resistors in parallel

When two elements connect at a single node pair, they are considered to be **in parallel**. Parallel-connected circuit elements have the same voltage across their terminals. The circuit shown in Figure 2.1.5 illustrates resistors connected in parallel. Don't make the mistake of assuming that two elements are parallel connected merely because they are lined up in parallel in circuit diagram. The defined characteristic of parallel-connected element is that they have the same voltage across their terminals.

In Figure 2.1.6, you can see that R_1 and R_3 are not parallel connected, because between their respective terminals, another resistor dissipates some of the voltage.

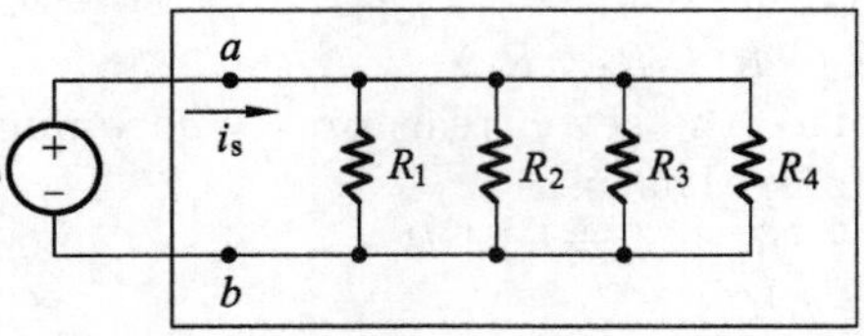

Figure 2.1.5 Resistor in parallel

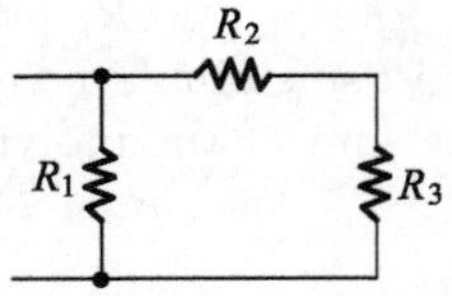

Figure 2.1.6 Nonparallel resistors

Resistors in parallel can be reduced to a single equivalent resistor using KCL and Ohm's law, as we now demonstrate. In the circuit shown in Figure 2.1.5, we let the currents i_1, i_2, i_3, and i_4 be the currents in the resistors R_1 through R_4 respectively. We also let the positive reference direction for each resistor current be down through the resistor, that is, from node a to node b using KCL:

$$i_s = i_1 + i_2 + i_3 + i_4 \tag{2.1.6}$$

The parallel connection of the resistors means that the voltage across each resistor must be the same. From Ohm's law,

$$i_1 R_1 = i_2 R_2 = i_3 R_3 = i_4 R_4 = v_s \tag{2.1.7}$$

therefore,

$$i_1 = \frac{v_s}{R_1}$$
$$i_2 = \frac{v_s}{R_2} \tag{2.1.8}$$
$$i_3 = \frac{v_s}{R_3}$$
$$i_4 = \frac{v_s}{R_4}$$

Substituting Equation (2.1.8) into Equation (2.1.6) yields

$$i_s = v_s\left(\frac{1}{R_1} + \frac{1}{R_2} + \frac{1}{R_3} + \frac{1}{R_4}\right) \tag{2.1.9}$$

from which

$$\frac{i_s}{v_s} = \frac{1}{R_{eq}} = \frac{1}{R_1} + \frac{1}{R_2} + \frac{1}{R_3} + \frac{1}{R_4} \tag{2.1.10}$$

For k resistors connected in parallel, Equation (2.1.10) becomes

$$\frac{1}{R_{eq}} = \sum_{i=1}^{k} \frac{1}{R_i} = \frac{1}{R_1} + \frac{1}{R_2} + \cdots + \frac{1}{R_k} \tag{2.1.11}$$

Note that the resistance of the equivalent resistor is always smaller than the resistance of the smallest resistor in the parallel connection. Sometimes, using conductance when resistors connected in parallel is more convenient. In that case, Equation (2.1.11) becomes

$$G_{eq} = \sum_{i=1}^{k} G_i = G_1 + G_2 + \cdots + G_k \tag{2.1.12}$$

2.1.3 Series versus parallel circuits

An element such as a toaster or a light bulb that absorbs power is called a **load.**

When we want to distribute power from a single voltage source to various loads, we usually place the loads in parallel. A switch in series with each load can break the flow of current to that load without affecting the voltage supplied to the other loads.

Sometimes, bulbs tend to fail or "burn out" by becoming open circuits. Then the entire string is dark and defective bulbs can be found only by trying each in turn. If several bulbs are burned out, it can be very tedious to locate the failed units. In a parallel connection, only the failed bulbs are dark.

2.1.4 Network analysis by using series and parallel equivalents

An electrical network (or electrical circuit) consists of circuit elements, such as resistance, voltage sources, and current sources, connected together to form closed paths. Network analysis is the process of determining the current, voltage, and power for each element, given the circuit diagram and the element values.

Sometimes, we can determine the current and voltages for each element in a resistive circuit by repeatedly replacing series and parallel combinations of resistances by their equivalent resistances. Eventually, this may reduce the circuit sufficiently that the equivalent circuit can be solved easily. The information gained from the simplified circuit is transferred to the precious steps in the chain of equivalent circuits. In the end, we gain enough information about the original circuit to determine all the currents and voltages.

2.1.5 Circuit analysis using series/parallel equivalents

Here are the steps in solving circuits using series/parallel equivalents:

(1) Begin by locating a combination of resistances that are in series or parallel. Often the place to start is the farthest from the source.

(2) Redraw the circuit with the equivalent resistance for the combination found in step 1.

(3) Repeat step 1 and step 2 until the circuit is reduced as far as possible. Often (but not always) we end up with a single source and a single resistance.

(4) Solve for the currents and voltages in the final equivalent circuit. Then, transfer results back one step and solve for additional unknown currents and voltages. Again transfer the results back one step and solve. Repeat until all of the currents and voltages are known in the original circuit.

(5) Check your results to make sure that KCL is satisfied at each node, KVL is satisfied for each loop, and the power adds to zero.

Example 2.1.1

Find i_1 and i_2 in the circuit shown in Figure A for example 2.1.1.

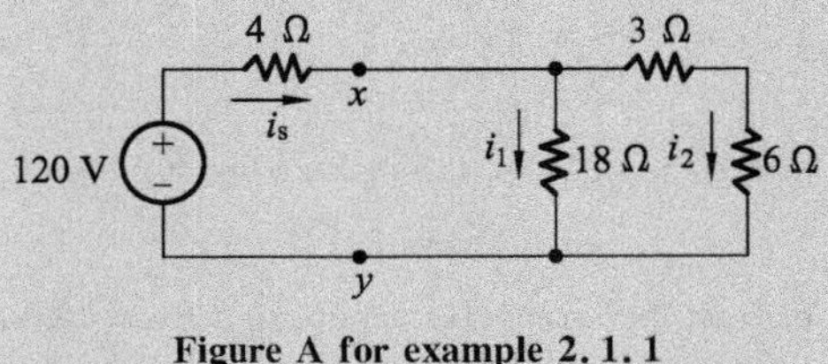

Figure A for example 2.1.1

Solution: We begin by noting that the 3 Ω resistor is in series with 6 Ω resistor. $R_{eq1}=3+6=9\ \Omega$

We replace this series combination with a 9 Ω resistor, reducing the circuit to the one shown in Figure B for example 2.1.1(a).

Now we can replace the parallel combination the 9 Ω and 18 Ω resistor with a single resistance 6 Ω.

$$R_{eq2}=\frac{18\times9}{18+9}=6\ \Omega$$

Figure B for example 2.1.1(b) shows this further reduction of the circuit.

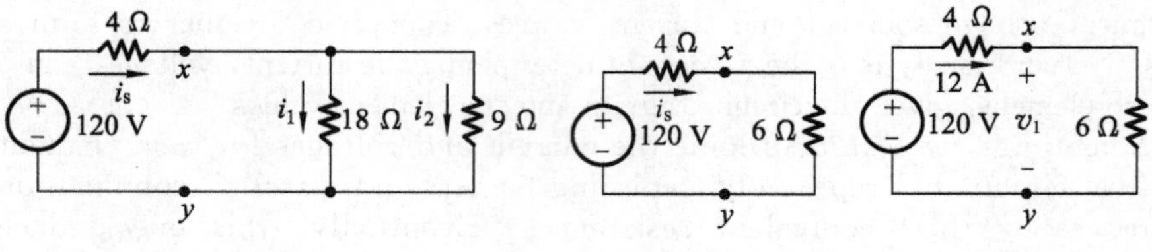

(a) (b) (c)

Figure B for example 2.1.1

From Figure B for example 2.1.1(b) we can verify that i_s equals 12 A.

$$i_s=\frac{120}{4+6}=10\ \text{A}$$

Based on Ohm's law we compute the value of v_1.

$$v_1=12\times6=72\ \text{V}$$

But v_1 is the voltage drop from node x to node y, so we can return to the circuit shown in Figure B for example 2.1.1(a),

$$i_1=\frac{v_1}{18}=\frac{72}{18}=4\ \text{A},\ i_2=\frac{v_1}{9}=\frac{72}{9}=8\ \text{A}$$

We can found the three specified currents by using series-parallel reduction in combination with Ohm's law.

2.2 The voltage-divider and current-divider circuits

2.2.1 Voltage-divider circuit

When a voltage is applied to a series combination of resistances, a fraction of the voltage appears across each of the resistance. Consider the circuit shown in Figure 2.2.1.

(a) A voltage-divider circuit (b) the voltage-divider circuit with current i indicated

Figure 2.2.1

We analyze this circuit by directly applying Ohm's law and Kirchhoff's law. To aid the analysis, we introduce the current i as shown in Figure 2.2.1(b). From KCL, R_1 and R_2 carry the same current. Applying KVL around the closed loop yields

$$v_s = iR_1 + iR_2 \tag{2.2.1}$$

or

$$i = \frac{v_s}{R_1 + R_2} \tag{2.2.2}$$

Now we can use Ohm's law to calculate v_1 and v_2:

$$v_1 = iR_1 = \frac{v_s R_1}{R_1 + R_2} \tag{2.2.3}$$

$$v_2 = iR_2 = \frac{v_s R_2}{R_1 + R_2} \tag{2.2.4}$$

Equations (2.2.3) and (2.2.4) show that v_1 and v_2 are fractions of v_s. Each fraction is the ratio of the resistance across which the divide voltage is defined to the sum of the two resistances. Because this ratio is always less than 1, the divided voltages v_1 and v_2 are always less than the source voltage v_s.

Example 2.2.1

Find the voltage v_0, in Figure for example 2.2.1.

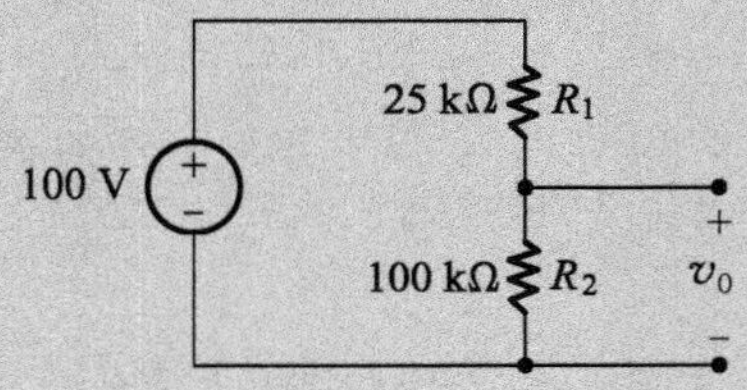

Figure for example 2.2.1

Solution:

$$v_0 = \frac{100}{25+100} \times 100 = 80\ \text{V}$$

2.2.2 Current-divider circuit

The current-divider circuit shown in Figure 2.2.2 consists of two resistors connected in parallel across a current source. The current-divider is designed to divide the current i_s between R_1 and R_2. We find the relationship between the current i_s and the current in each resistor (that is i_1 and i_2) by directly applying Ohm's law and KCL.

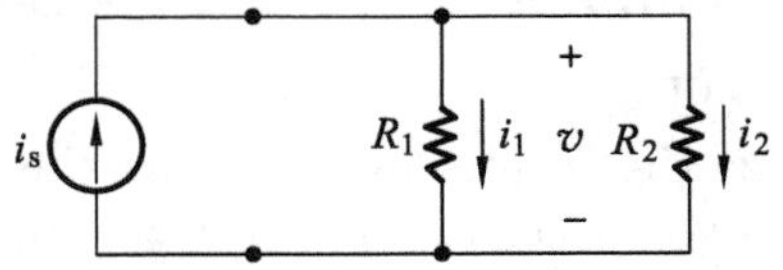

Figure 2.2.2 The current-divider circuit

The voltage across the parallel resistor is

$$v = i_1 R_1 = i_2 R_2 = \frac{R_1 R_2}{R_1 + R_2} i_s \tag{2.2.5}$$

From Equation (2.2.5)

$$i_1=\frac{R_2}{R_1+R_2}i_s \tag{2.2.6}$$

$$i_2=\frac{R_1}{R_1+R_2}i_s \tag{2.2.7}$$

Equations (2.2.6) and (2.2.7) show that the current divided between two resistors in parallel so that the current in one resistor equals the current entering the parallel pair multiplied by the other resistance and divided by the sum of the resistors.

Example 2.2.2

Find the power dissipated in the 6 Ω resistor shown in Figure for example 2.2.2.

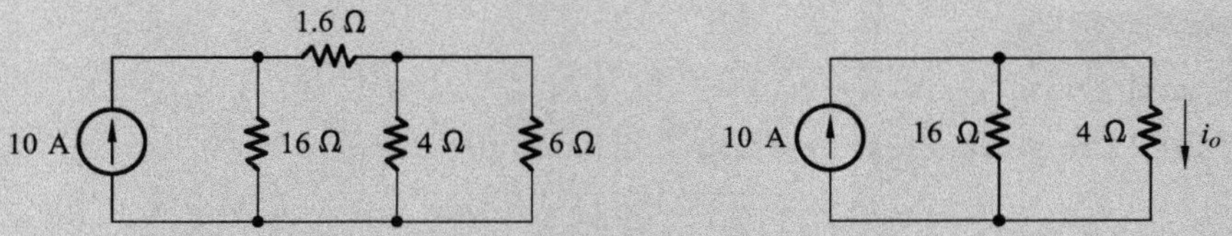

(a) The circuit (b) A simplification of the circuit for example 2.2.2

Figure for example 2.2.2

Solution: First, we must find the current in the resistor by simplfying the circuit with series-parallel reductions.

$$R_{eq}=1.6+\frac{4\times 6}{4+6}=4\ \Omega$$

We can find current i_0 by using the formula for current division:

$$i_0=\frac{16}{16+4}\times 10=8\ \text{A}$$

We now can further divide i_0 between the 6 Ω and 4 Ω resistors.

$$i_6=\frac{4}{6+4}\times 8=3.2\ \text{A}$$

And the power dissipated in the 6 Ω resistor is

$$p=(3.2)^2\times 6=61.44\ \text{W}$$

2.3 Voltage division and current division

2.3.1 Voltage division

We can now generalize the results by analyzing the voltage-divider circuit in Figure 2.2.1 and current-divider circuit in Figure 2.2.2. The generalizations will yield two additional and very useful circuit analysis techniques known as voltage division and current division. Consider the circuit shown in Figure 2.3.1.

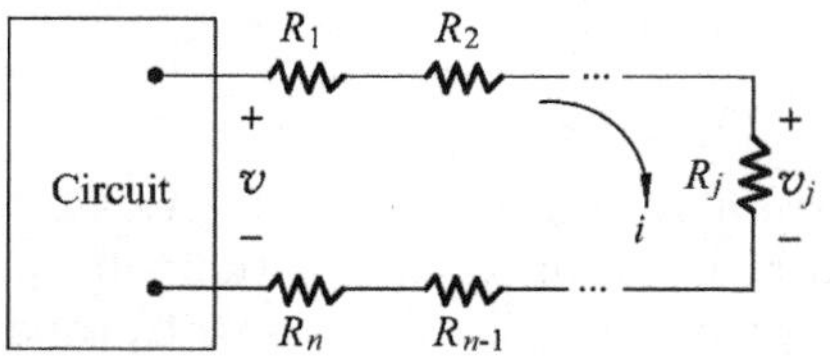

Figure 2.3.1 Circuit used to illustrate voltage division

The box on the left of Figure 2.3.1 contains a single voltage source or any other combination of basic circuit elements that results in the voltage v shown in the figure. On the right of the box are n resistors connected in series. We are interested in finding the voltage drop v_j across an arbitrary resistor R_j in terms of the voltage v. We start by using Ohm's law to calculate i, the current through all of the resistors in series, in terms of the voltage v and the n resistors:

$$i=\frac{v}{R_1+R_2+\cdots+R_n}=\frac{v}{R_{eq}} \tag{2.3.1}$$

The equivalent resistance R_{eq} is the sum of the n resistor values because the resistors are in series, as shown in Equation (2.1.3). We apply Ohm's law to calculate the voltage drop v_j across the resistor R_j using the current i calculated in Equation (2.3.1):

$$v_j=iR_j=\frac{R_j}{R_{eq}}v \tag{2.3.2}$$

This is the voltage division equation. It states that the voltage drop v_j across a single resistor R_j from a collection of series-connected resistors is proportional to the total voltage drop v across the set of series-resistors. The constant of proportionality is the ratio of the single resistance to the equivalent resistance of the series connected set of resistors, or $\frac{R_j}{R_{eq}}$.

2.3.2 Current division

The box on the left can contain a single current source or any other combination of basic circuit elements that results in current i shown in Figure 2.3.2.

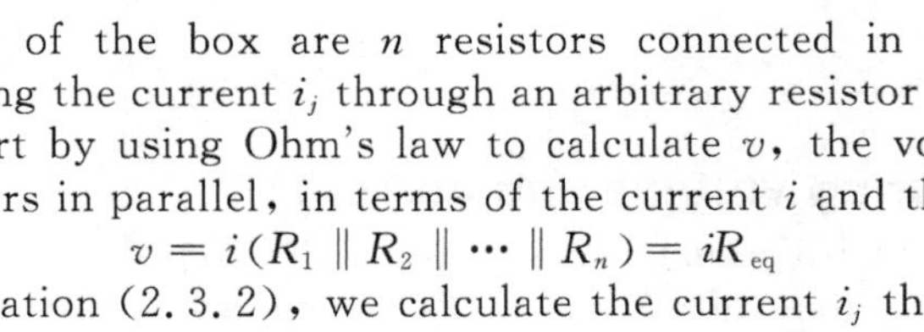

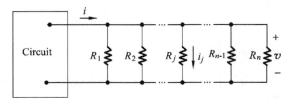

Figure 2.3.2 Circuit used to illustrate current division

To the right of the box are n resistors connected in parallel. We are interested in finding the current i_j through an arbitrary resistor R_j in terms of the current i. We start by using Ohm's law to calculate v, the voltage drop across each of the resistors in parallel, in terms of the current i and the n resistors.

$$v = i(R_1 \parallel R_2 \parallel \cdots \parallel R_n)= iR_{eq} \tag{2.3.3}$$

Based on Equation (2.3.2), we calculate the current i_j through the resistor

R_j, using the voltage v calculated in Equation (2.3.3)

$$i_j = \frac{v}{R_j} = \frac{R_{eq}}{R_j} i \tag{2.3.4}$$

It says that the current i_j through a single resistor R_j from a collection of parallel-connected resistors is proportional to the total current i supplied to the set of parallel-connected resistors. The constant of proportionality is the ratio of the single resistance to the equivalent resistance of the series connected set of resistors, or R_{eq}/R_j. Note that the constant of proportionality in the current division equation is the inverse of the constant of proportionality in the voltage division equation.

Example 2.3.1

Use current division to find the current i_0 and use voltage division to find the voltage v_0 for the circuit in Figure for example 2.3.1.

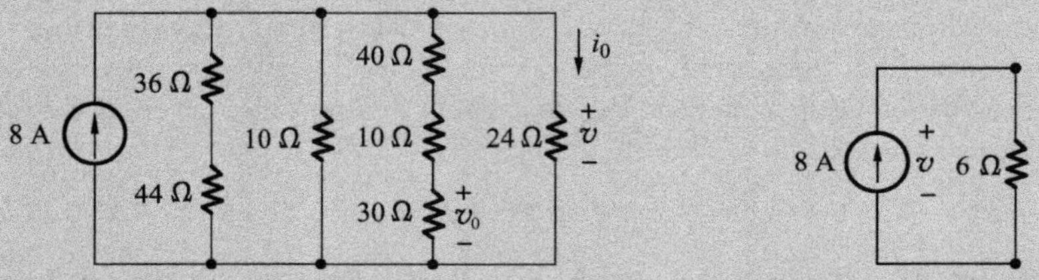

(a) The circuit

(b) A simplification of the circuit for example 2.3.1

Figure for example 2.3.1

Solution: We can find the equivalent resistance of the four parallel branches containing resistors.

$$\begin{aligned} R_{eq} &= 24 \parallel (40+10+30) \parallel 10 \parallel (36+44) \\ &= 24 \parallel 80 \parallel 10 \parallel 80 = 6\ \Omega \end{aligned}$$

Applying the Equation (2.3.4),

$$i_0 = \frac{6}{24} \times 8 = 2\ \text{A}$$

we can use Ohm's law to find v,

$$v = 6 \times 8 = 48\ \text{V}$$

This is also the voltage drop across the branch containing the 40 Ω, 10 Ω and 30 Ω resistors in series.

Then we can use voltage division to determine the voltage drop v_0 across the 30 Ω:

$$v_0 = \frac{30}{40+10+30} \times 48 = 18\ \text{V}$$

2.4 Measuring voltage and current

When working with actual circuits, you will often need to measure voltages and currents. We will spend some time discussing several measuring devices here and in the next section, because they are relatively simple to analyze and offer practical examples of the current-and voltage-divider configurations we have just studied.

An **ammeter** is an instrument designed to measure current; it is placed in series with the circuit element whose current is being measured. A **voltmeter** is an instrument designed to measure voltage; it is placed in parallel with the circuit element whose voltage is being measured. An ideal ammeter or voltmeter has no effect on the circuit variable it is designed to measure. That is, an ideal ammeter has an equivalent resistance of 0 Ω and functions as a short circuit in series with the element whose current is being measured. An ideal voltmeter has an infinite equivalent resistance and thus functions as an open circuit in parallel with the element whose voltage is being measured. The configuration for an ammeter used to measure the current in R_1 and for a voltmeter used to measure the voltage in R_2 is depicted in Figure 2. 4. 1. The ideal model for these meters in the same circuit is shown in Figure 2. 4. 2.

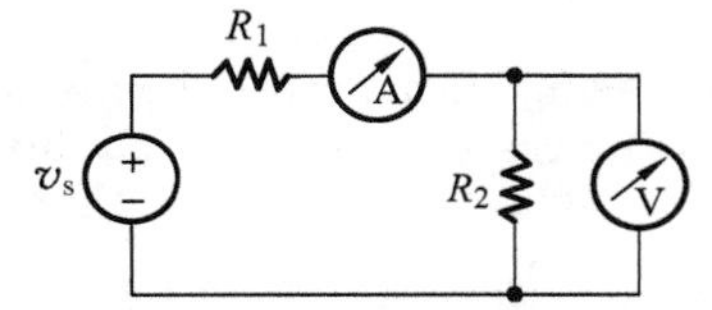

Figure 2. 4. 1 An ammeter connected to measure the current in R_1, and a voltmeter connected to measure the voltage across R_2

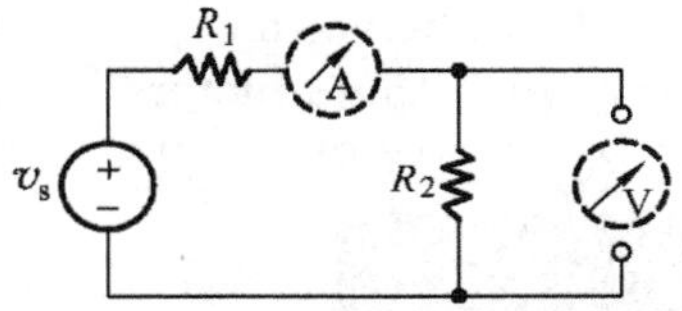

Figure 2. 4. 2 A shout-circuit model for the ideal ammeter, and an open-circuit model for the ideal voltmeter

There are two broad categories of meters used to measure continuous voltages and currents: digital meters and analog meters. **Digital maters** measure the continuous voltage or current signal at discrete points in time, called the sampling times. The signal is thus converted from an analog signal, which is continuous in time, to a digital signal, which exists only at discrete instants in time.

Analog meters are based on the D'Arsonval meter movement which implements the readout mechanism. A D'Arsonval meter movement consists of a movable coil placed in the field of a permanent magnet. When current flows in the coil, it creates a torque on the coil, causing it to rotate and move a pointer across a calibrated scale. By design, the deflection of the pointer is directly proportional to the current in the movable coil. The coil is characterized by both a voltage rating and a current rating.

An analog ammeter consists of a D'Arsonval movement in parallel with a resistor, as shown in Figure 2. 4. 3. The purpose of the parallel resistor is to limit the amount of current in the movement's coil by shunting some of it through R_A. An analog voltmeter consists of a D'Arsonval movement in series with a resistor, as shown in Figure 2. 4. 4. Here, the resistor is used to limit the voltage drop across the meter's coil. In both meters, the added resistor determines the full-scale reading of the meter movement.

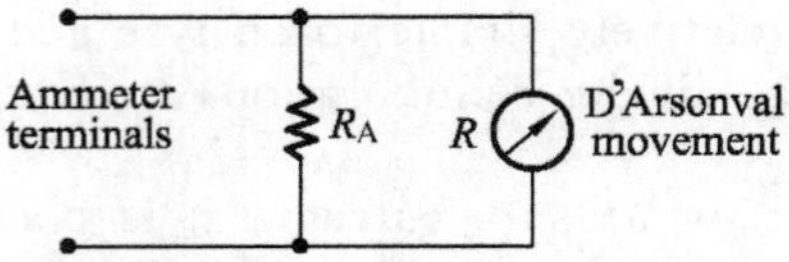

Figure 2.4.3 A DC ammeter circuit

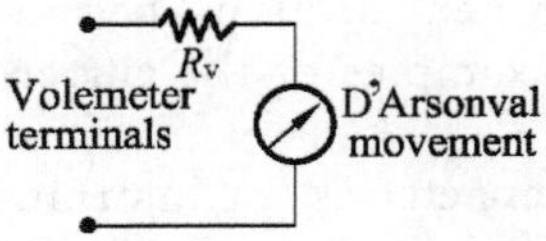

Figure 2.4.4 A DC voltmeter circuit

From these descriptions we see that an actual meter is non-ideal; both the added resistor and the meter movement introduce resistance in the circuit to which the meter is attached. In fact, any instrument used to make physical measurement extracts energy from the system while making measurements. The more energy is extracted by the instruments, the more severely the measurement is disturbed. A real ammeter has an equivalent resistance that is not zero, and it thus effectively adds resistance to the circuit in series with the element whose current the ammeter is reading. A real voltmeter has an equivalent resistance that is not infinite, so it effectively adds resistance to the circuit in parallel with the element whose voltage the voltmeter is being read.

How much these meters disturb the circuit being measured depends on the effective resistance of the meters compared with the resistance in the circuit.

Example 2.4.1

(a) A 50 mV, 1 mA D'Arsonval movement is to be used in an ammeter with a full-scale reading of 10 mA. Determine R_A.

(b) Repeat (a) for a full-scale reading of 1 A.

(c) How much resistance is added to the circuit when the 10 mA ammeter is inserted to measure current?

(d) Repeat (c) for the 1 A ammeter.

Solution: (a) From the statement of the problem, we know that when the current at the terminals of the ammeter is 10 mA, 1 mA is flowing through the meter coil, which means that 9 mA must be diverted through R_A. We also know that when the movement carries 1 mA, the drop across its terminals is 50 mV. Ohm's law requires that

$$9\times10^{-3}R_A=50\times10^{-3}$$

$$\therefore\ R_A=\frac{50}{9}=5.555\ \Omega$$

(b) When the full-scale deflection of the ammeter is 1 A, R_A must carry 999 mA when the movement carries 1 mA. In this case, then,

$$999\times10^{-3}R_A=50\times10^{-3}$$

$$\therefore\ R_A=\frac{50}{999}=50.05\ \text{m}\Omega$$

(c) Let R_m represent the equivalent resistance of the ammeter. For the 10 mA ammeter,

$$R_m=\frac{50\ \text{mV}}{10\ \text{mA}}=5\ \Omega$$

Or, alternatively,
$$R_m=\frac{(50)\left(\frac{50}{9}\right)}{50+\left(\frac{50}{9}\right)}=5\ \Omega$$

(d) For the 1 A ammeter
$$R_m=\frac{50\ \text{mV}}{1\ \text{A}}=0.050\ \Omega$$

Or, alternatively
$$R_m=\frac{(50)\left(\frac{50}{999}\right)}{50+\left(\frac{50}{999}\right)}=0.050\ \Omega$$

Example 2.4.2

(a) A 50 mV, 1 mA D'Arsonval movement is to be used in a voltmeter with a full-scale reading of 150 V. Determine R_v.

(b) Repeat (a) for a full-scale reading of 5 V.

(c) How much resistance does the 150 V meter insert into the circuit?

(d) Repeat (c) for the 5 V meter.

Solution: (a) Full-scale deflection requires 50 mV across the meter movement, and the movement has a resistor of 50 Ω. Therefore, we apply Equation (2.2.4) with: $R_1=R_v$, $R_2=50\ \Omega$, $v_s=150$ V, $v_2=50$ mV:

$$50\times10^{-3}=\frac{50}{R_v+50}\times150$$

Solving for R_v gives:

$$R_v=149\ 950\ \Omega$$

(b) For a full-scale reading of 5 V,

$$50\times10^{-3}=\frac{50}{R_v+50}\times5$$

$$R_v=4\ 950\ \Omega$$

(c) If we let R_m represent the equivalent resistance of the meter,

$$R_m=\frac{150\ \text{V}}{10^{-3}\ \text{A}}=150\ 000\ \Omega$$

or
$$R_m=149\ 950+50=150\ 000\ \Omega$$

(d) Then,
$$R_m=\frac{5\ \text{V}}{10^{-3}\ \text{A}}=5\ 000\ \Omega$$

or
$$R_m=4\ 950+50=5\ 000\ \Omega$$

2.5 Measuring resistance—the Wheatstone bridge

The **Wheatstone bridge** is a circuit used to measure unknown resistances. For example, it is used by mechanical and civil engineers to measure the resistances of strain gauges in experimental stress studies of machines and building. The circuit is shown in Figure 2.5.1. The circuit consists of a DC source v_s, a detector, the unknown resistance to be measured R_x, and three precision resistors, R_1, R_2

and R_3. Usually, R_2 and R_3 are adjustable resistances, which are indicated in the figure by the arrow drawn through the resistance symbols.

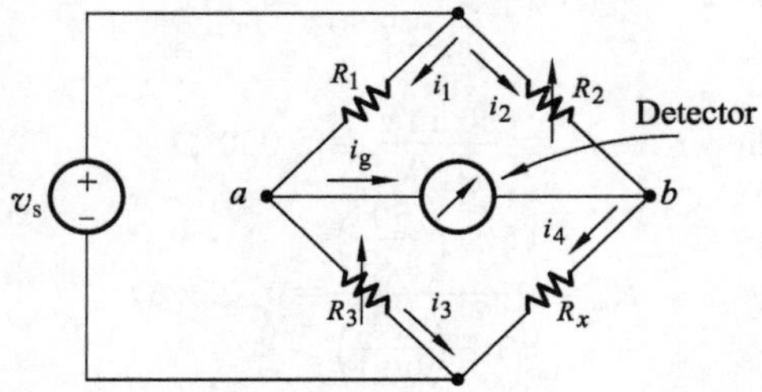

Figure 2. 5. 1 The Wheatstone bridge

The detector is capable of responding to very small currents (less than 1 mA). However, it is not necessary for the detector to be calibrated. It is only necessary for the detector to indicate whether or not current is flowing through it. Often, the detector is a D'Arsonval galvanometer, which has a pointer that deflects one way or the other, depending on the direction of the current through it.

In operation, the resistors R_2 and R_3 are adjusted in value until the detector indicates zero current. In this condition, we say that the bridge is balanced. Then, the current i_g and the voltage across the detector v_{ab} are zero.

Applying KCL at node a (Figure 2. 5. 1) and using the fact that $i_g=0$, we have

$$i_1=i_3 \tag{2. 5. 1}$$

Similarly, at node b, we get

$$i_2=i_4 \tag{2. 5. 2}$$

Writing a KVL equation around the loop formed by R_1, R_2, and the detector, we obtain

$$R_1 i_1+v_{ab}=R_2 i_2 \tag{2. 5. 3}$$

However, when the bridge is balanced, $v_{ab}=0$, so that

$$R_1 i_1=R_2 i_2 \tag{2. 5. 4}$$

Similarly, for the loop consisting of R_3, R_4 and the detector, we have

$$R_3 i_3=R_x i_4 \tag{2. 5. 5}$$

Using Equations (2. 5. 1) and (2. 5. 2) to the substitute into Equation (2. 5. 5), we obtain

$$R_3 i_3=R_x i_2 \tag{2. 5. 6}$$

Through the Equations (2. 5. 4) and (2. 5. 6), we get

$$R_x=R_2 R_3/R_1 \tag{2. 5. 7}$$

Often, in commercial bridge, a multiposition switch selects an order-of-magnitude scale factor R_2/R_1 by changing the value of R_2. Then, R_3 is adjusted by means of calibrated switches until balance is achieved. Finally, the unknown resistance R_x is the scale factor times the value of R_3.

Summary

The objective of this chapter is to introduce the basic analysis of linear resistive circuits.

1. Understand the connection of series and parallel circuits, knowing how to find the equivalent resistance.

2. Apply the voltage and current division laws to calculate unknown variables in simple circuit.

3. Understand the rules for connecting electric measuring instruments to electric circuits for the measurement of voltage, current, and power.

Problems

2.1 For each of circuits shown in Figure for problem 2.1, find

(a) The equivalent resistance seen by the source.

(b) The power developed by the source.

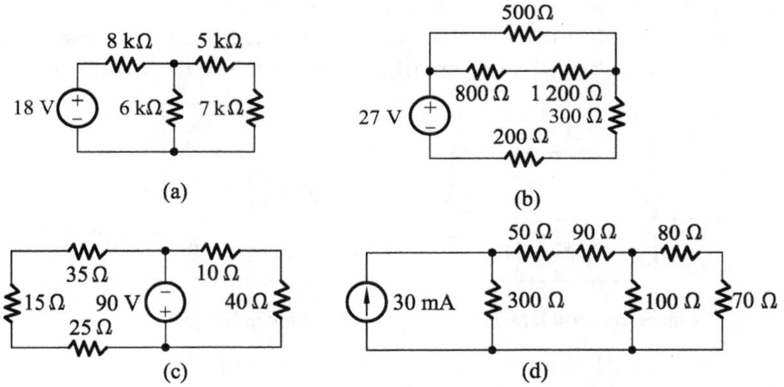

Figure for problem 2.1

2.2 For each of circuits shown in Figure for problem 2.2.

(a) Find the power dissipated in each resistor.

(b) Find the power delivered by the 120 V source.

(c) Show that the power delivered equals the power dissipated.

2.3 For the circuit shown in Figure for problem 2.3, find

(a) Voltage v.

(b) The power delivered to the circuit by the current source.

(c) The power dissipated in the 10 Ω resistor.

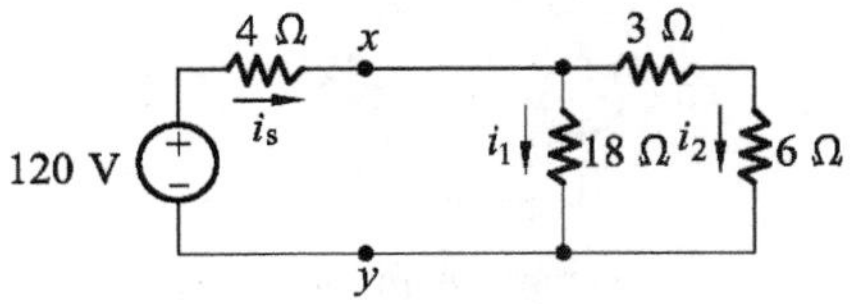

Figure for problem 2.2

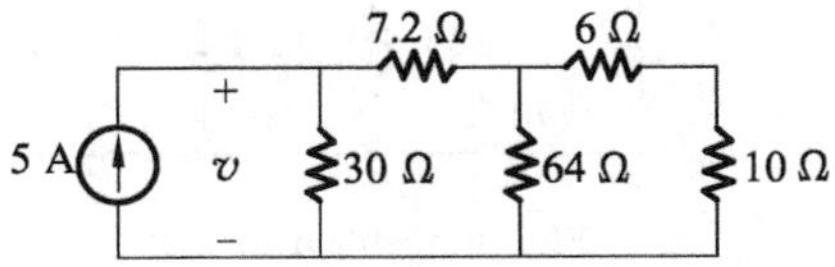

Figure for problem 2.3

2.4 As the circuit shown in Figure for problem 2.4

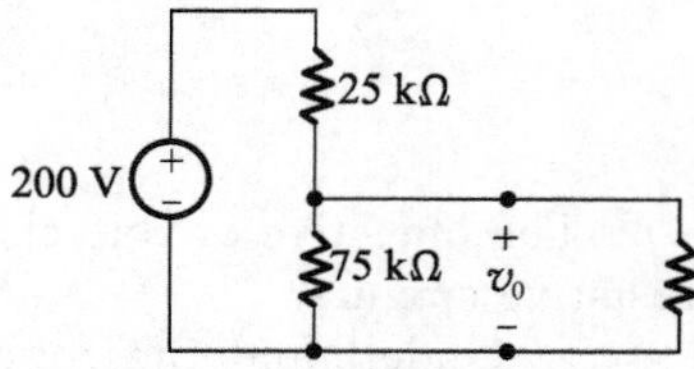

Figure for problem 2.4

(a) Find the no-load value of v_0.

(b) Find v_0 when R_L is 150 kΩ.

(c) How much power is dissipated in the 25 kΩ resistor if the load terminals are accidentally short-circuited.

(d) What is the maximum power dissipated in the 75 kΩ resistor?

2.5 As the circuit shown in Figure for problem 2.5.

(a) Find the value of R that will cause 4 A of current to flow through the 80 Ω resistor.

(b) How much power will the resistor R from part (a) need to dissipate?

(c) How much power will the current source generate for the value of R from part (a)?

2.6 As the circuit shown in Figure for problem 2.6.

(a) Calculate the no-load voltage v_o for the voltage-divider.

(b) Calculate the power dissipated in R_1 and R_2.

(c) Assume that only 0.5 W resistor are available. The no-load voltage is to be the same as in (a). Specify the smallest Ohmic value of R_1 and R_2.

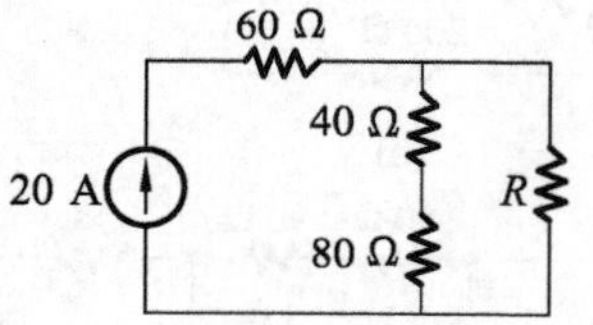

Figure for problem 2.5

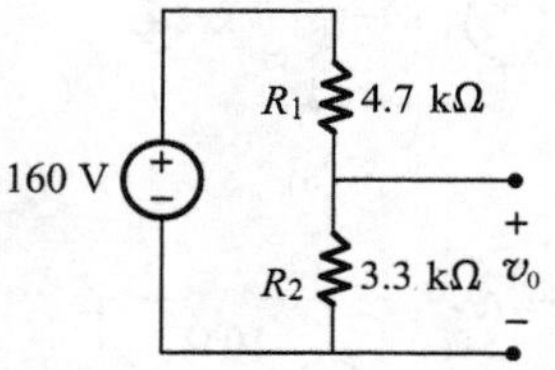

Figure for problem 2.6

2.7 As the circuit shown in Figure for problem 2.7.

(a) Use voltage division to determine the voltage v_0 across the 40 Ω resistor.

(b) Use v_0 from part (a) to determine the current through the 40 Ω resistor, and use this current and current division to calculate the current in the 30 Ω resistor.

(c) How much power is absorbed by the 50 Ω resistor?

2.8 The no-load voltage in the voltage-divider circuit shown in Figure for problem 2.8 is 8 V. The smallest load resistor that is ever connected to the divider is 3.6 kΩ. When the divider is loaded, v_0 is not to drop below 7.5 V.

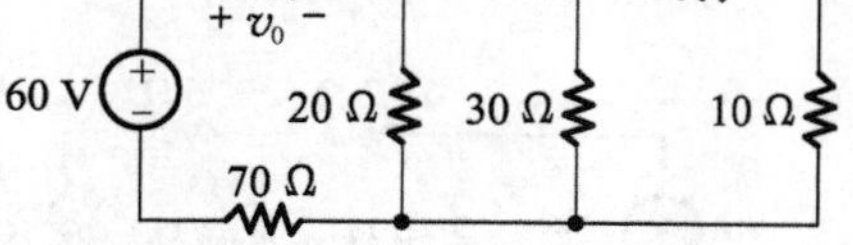

Figure problem for 2.7

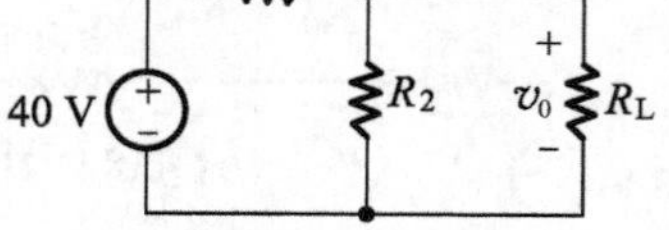

Figure for problem 2.8

(a) Design the divider circuit to meet the specifications just mentioned.

Specify the numerical values of R_1 and R_2.

(b) Assume the power ratings of commercially available resistors are $\frac{1}{16}$, $\frac{1}{8}$, $\frac{1}{4}$ and 2 W. What power rating would you specify?

2.9 Attach a 450 mA current source between the terminals *a-b* in Figure for problem 2.9, with the current arrow pointing up.

(a) Use current division to find the current in the 36 Ω resistor from top to bottom.

(b) Use the result from part (a) to find the voltage across the 36 Ω resistor, positive at the top.

(c) Use the result from part (b) and voltage division to find the voltage across the 18 Ω resistor, positive at the top.

(d) Use the result from part (c) and voltage division to find the voltage across the 10 Ω resistor, positive at the top.

2.10 As the circuit shown in Figure for problem 2.10.

(a) Find the current.

(b) If the ammeter in Example 2.4.1 is used to measure the current, what will it read?

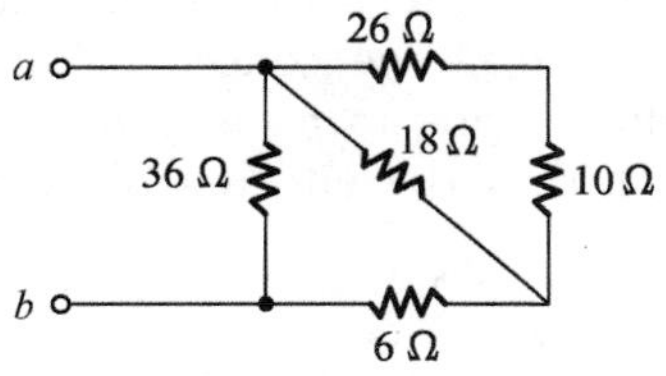

Figure for problem 2.9

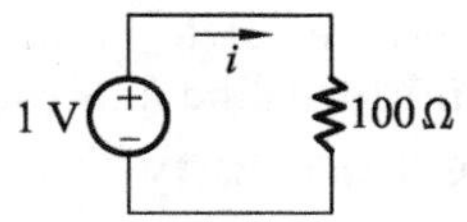

Figure for problem 2.10

2.11 As the circuit shown in Figure for problem 2.11.

(a) Find the voltage v across the 75 kΩ resistor.

(b) If the 150 V volmeter of Example 2.4.2(a) is used to measure the voltage, what will be the reading?

2.12 The bridge circuit shown in Figure for problem 2.12 is balanced when $R_1 = 100\ \Omega$, $R_2 = 1\ 000\ \Omega$, $R_3 = 150\ \Omega$. The bridge is energized from a 5 V DC source.

(a) What is the value of R_x?

(b) Suppose each bridge resistor is capable of dissipating 250 mW. Can the bridge be left in the balanced state without exceeding the power-dissipating capacity of the resistors, thereby damaging the bridge?

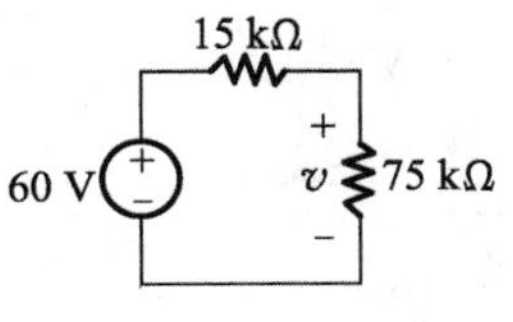

Figure for problem 2.11

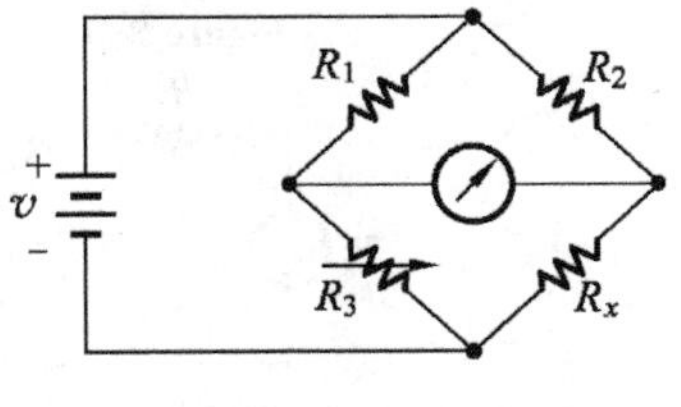

Figure for problem 2.12

CHAPTER 3
TECHNIQUES OF CIRCUIT ANALYSIS

Introduction

This chapter illustrates the fundamental techniques for the analysis of resistive circuits. The chapter begins with the definition of network variables and network analysis problems. Then, the two most widely applied methods—node analysis and mesh analysis—are introduced. These are the most generally applicable circuit solution techniques used to derive the equations of all electric circuits; their application to resistive circuits in this chapter is intended to acquaint you with these methods, which are used throughout the book. The second solution method presented is based on the principle of superposition, which is applicable only to liner circuits. Next, the concept of Thevenin and Norton equivalent circuits is explored, which leads to a discussion of maximum power transfer in electric circuits and facilitates the ensuing discussion of non-linear loads and load-line analysis.

3.1 Terminology

To discuss the more involved methods of circuit analysis, we must define a few basic terms. So far, all circuits that can be drawn on a plane without having one element (or conductor) crossing over another are called planar circuits. Figure 3. 1. 1 is a **planar circuit.** On the other hand, circuits that must be drawn with one or more elements crossing others are said to be non-planar circuit, like Figure 3. 1. 2.

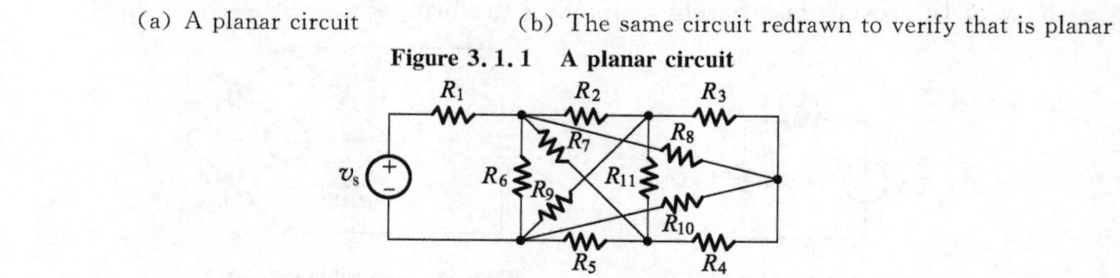

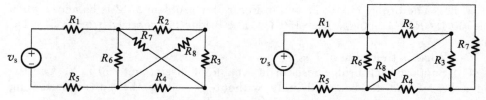

(a) A planar circuit (b) The same circuit redrawn to verify that is planar

Figure 3. 1. 1 A planar circuit

Figure 3. 1. 2 A non-planar circuit

We consider only when basic circuit elements are interconnected to form a circuit, the resulting interconnection is described in terms of nodes, paths, branches, loops, meshes, etc.. All of these definitions are presented in Table 3. 1. 1. Table 3. 1. 1 also includes examples of each definition taken from the circuit in Figure 3. 1. 3.

Table 3. 1. 1 Terms for describing circuits

name	definition	Example from Figure 3. 1. 3
Node	A point where two or more circuit elements join	a
Essential node	A point where three or more circuit elements join	b
Path	A trace of adjoining basic elements with no elements included more than once	$v_1-R_1-R_5-R_6$
Branch	A path that connects two nodes	R_1
Essential branch	A path that connects two essential nodes without passing through an essential node	v_1-R_1
Loop	A path whose last node is the same as the starting	$v_1-R_1-R_5-R_6-R_4-v_2$
Mesh	A loop that does not enclose any other loop	$v_1-R_1-R_5-R_3-R_2$
Planar circuit	A circuit that can be drawn on a plane with no crossing branches	Figure 3. 1. 1 is a planar circuit Figure 3. 1. 2 is a non-planar circuit

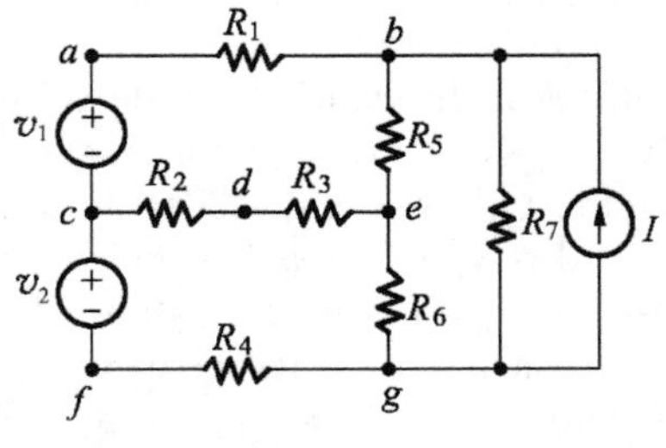

Figure 3. 1. 3 A circuit illustrating nodes, branches, meshes, paths, and loops

3.2 Node-voltage method

The network analysis methods that we have studied so far are useful, but they do not apply to all networks. For example, consider the circuit shown in Figure 3. 2. 1. We cannot solve this circuit by combining resistances in series or parallel because no series or parallel combination of resistances exist in the circuit. Furthermore, the voltage-division and current-division principle cannot be applied to this circuit. In this section, we learn node-voltage analysis, which is a general technique that can be applied to any circuit.

3.2.1 Reference node

A **node** is a point at which two or more circuit elements are joined together. In node-voltage analysis, we first select one of the nodes as the **reference node.** In principle, any node can be picked to be the reference node. However, the solution is usually facilitated by selecting one end of a voltage sources as the reference node.

For example, the circuit shown in Figure 3.2.1 has 4 nodes. We select the bottom node as the reference node, and mark the reference node by the **ground symbol**, as shown in the Figure 3.2.1.

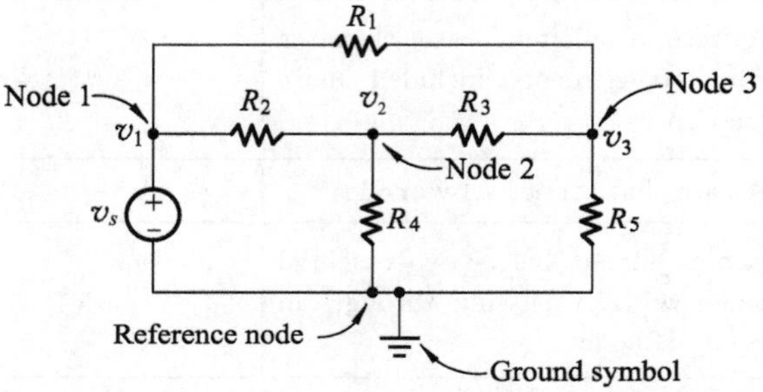

Figure 3.2.1 The first step in node-voltage analysis is to select a reference node and label the voltages at each of the other nodes.

3.2.2 Node voltages

Next, we label the voltages at each of the other nodes. For example, the voltages at the 3 nodes are labeled v_1, v_2 and v_3 in Figure 3.2.1. The v_1 is the voltage between node 1 and the reference node. The reference polarity for v_1 is positive at node 1 and negative at the reference node. Similarly, v_2 is the voltage between node 2 and the reference node. The reference polarity for v_2 is positive at node 2 and negative at the reference node.

3.2.2.1 Element voltages in terms of the node voltages

In node-voltage analysis, we write equations and eventually solve for the node voltages. Once the node voltages have been found, it is relatively easy to find the current, voltage, and power for each element in the circuit.

For example, suppose that we know the values of the node voltages and we want to find the voltage across R_3 with its positive reference on the left-hand side. The node voltages and the voltage v_x across R_3 are shown in Figure 3.2.2, where arrows are used to indicate reference polarities. Notice that v_2, v_x and v_3 are the voltages encountered in traveling around the closed path through R_4, R_3, and R_5.

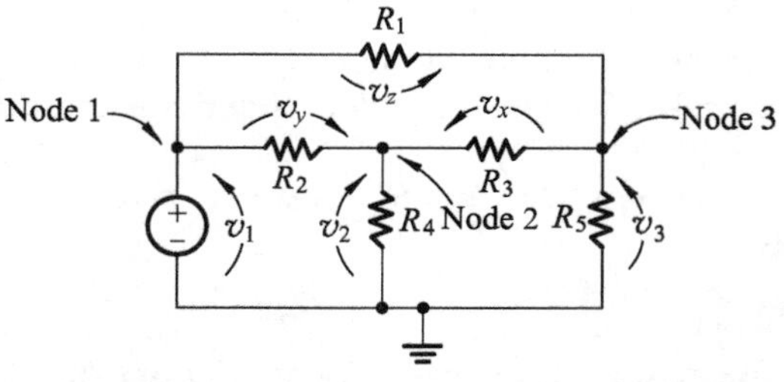

Figure 3.2.2 Assuming that we can determine the node voltages v_1, v_2, and v_3, we can use KVL to determine v_x, v_y, and v_z

Thus, these voltages must obey KVL. Traveling around the loop anti clockwise and summing voltages, we have

$$-v_2+v_x+v_3=0$$

Solving for v_x, we obtain

$$v_x=v_2-v_3$$

We can use the same way to find

$$v_y=v_2-v_1$$

$$v_z=v_3-v_1$$

3.2.2.2 KCL equations in terms of the node voltages

After choosing the reference node and assigning the voltage variables, we write equations that can be solved for the node voltages. We demonstrate by continuing with the circuit of Figure 3.2.2. In Figure 3.2.2, the voltage v_1 is the same as the source voltage v_s:

$$v_1=v_s$$

(In this case, one of the node voltages is known without any effort. This is the advantage in selecting the reference node at one end of an independent voltage source.)

Therefore, we need to determine the values of v_2 and v_3, and must write two independent equations. We usually start by trying to write current equations at each of the nodes corresponding to an unknown node voltage. At node 2 in Figure 3.2.2, the current leaving through R_4 is given by

$$V_2/R_4$$

Next, referring to Figure 3.2.2, we see that current flowing out of node 2 through R_3 is given by v_x/R_3.

However, we found earlier that $v_x=v_2-v_3$. Thus, the current flowing out of node 2 through R_3 is given by

$$(v_2-v_3)/R_3$$

At this point, we get the very useful observation: to find the current flowing out of node n though a resistance toward node k, we subtract the voltage at node k from the voltage at node n and divide the difference by

$$(v_n-v_k)/R$$

Applying this observation in Figure 3.2.2 to find the current flowing out of node 2 through R_2, we have

$$(v_2-v_1)/R_2$$

Now we apply KCL, adding all of the expressions for the currents leaving node 2 and setting the sum to zero. Thus, we obtain

$$(v_2-v_1)/R_2+V_2/R_4+(v_2-v_3)/R_3=0$$

Writing the current equation at node 3 is similar. We usually write expressions for the currents leaving the node under consideration and set the sum to zero.

$$(v_3-v_1)/R_1+V_3/R_5+(v_3-v_2)/R_3=0$$

Example 3.2.1

Solve for the node voltages shown in Figure A for example 3.2.1 and determine the value of the current i_x.

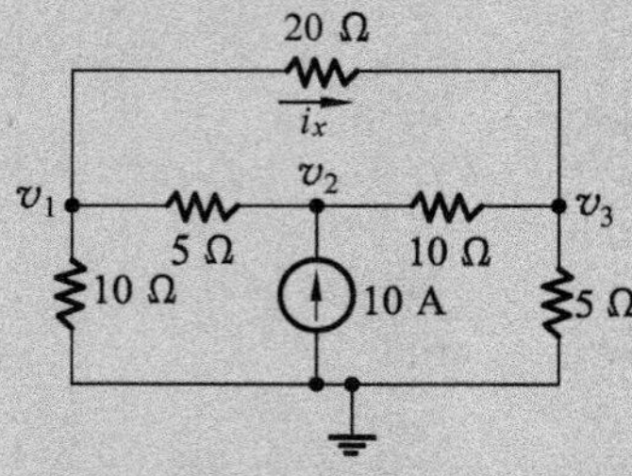

Figure A for example 3. 2. 1

Solution: First, we select the reference node and assign the node voltages. This has already been done, as shown in Figure for example 3. 2. 1.

Next, we write equations. In this case, we can write a current equation at each node. This yields

node 1: $v_1/10+(v_1-v_2)/5+(v_1-v_3)/20=0$

node 2: $(v_2-v_1)/5+(v_2-v_3)/10=10$

node 3: $v_3/5+(v_3-v_2)/10+(v_3-v_1)/20=0$

Finally, we solve this equation set, and we get $i_x=0.9091$ A

Repeat the analysis of the circuit of Example 3. 2. 1, using different reference nodes and node voltages shown in Figure B for example 3. 2. 1.

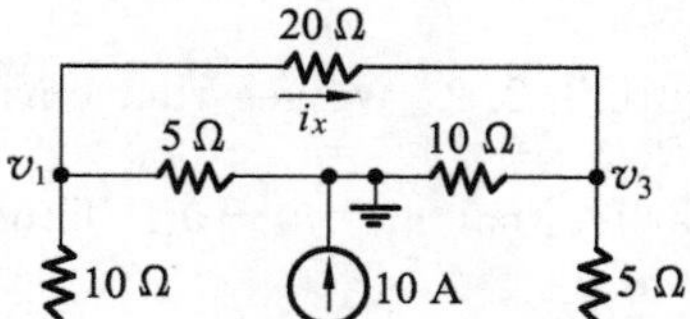

Figure B for example 3. 2. 1 with a different choice for the reference node.

We give equation set directly

$$(v_1-v_3)/20+v_1/5+(v_1-v_2)/10=0$$
$$(v_2-v_1)/10+10+(v_2-v_3)/5=0$$
$$(v_3-v_1)/20+v_3/10+(v_3-v_2)/5=0$$

Then we get the same result. Through this we know we can choose any node as a reference node, for it does not change the result.

3.2.3 Supernode

Another way to obtain a current equation is to form a **Supernode.** This is done by drawing a dashed line around several nodes, including the elements connected between them. This is shown in Figure 3. 2. 3. Two Supernodes are indicated, one enclosing each of the voltage sources.

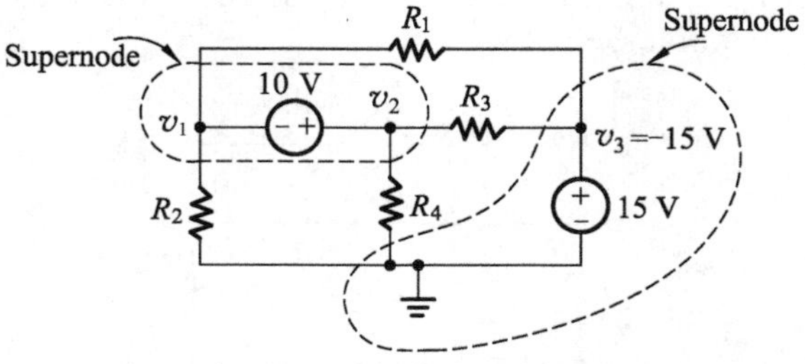

Figure 3. 2. 3 A Supernode is formed by drawing a dashed line around several nodes, including the elements connected between them

We can state KCL in a slightly more general form than we have previously: the current flowing through any closed surface must equal zero. Thus, we can apply KCL to a Supernode. For example, for the Supernode enclosing the 10 V source, we sum currents leaving and obtain

$$V_1/R_2+[v_1-(-15)]/R_1+v_2/R_4+[v_2-(-15)]/R_3=0 \qquad (3.2.1)$$

Each term on the left-hand side of this equation represents a current leaving the Supernode through one of the resistors. Thus, by enclosing the 10 V source within the Supernode, we have obtained a current equation without introduceing a new variable for the current in the source.

Next, we can write another current equation for the other Supernode.

$$V_1/R_2+(v_1+15)/R_1+v_2/R_4+(v_2+15)/R_3=0 \qquad (3.2.2)$$

Comparing Equations (3. 2. 1) and (3. 2. 2), we can find they are totally the same. Nodes 1 and 2 are part of the first Supernode, while node 3 and the reference node are part of the second Supernode. Thus, in writing equations for both Supernodes, we would have used all 4 nodes in the network. Actually, when we use KCL to write current equations, n nodes just have (n-1) independent equations (application of the current law to the nth node does not generate an independent equation, because this equation can be derived from the previous n-1 equations.). But n unknown quantities must have n equations, so that we can resolve the problem. So here we can use KVL because v_1, the 10 V source, and v_2 form a closed loop. Traveling clockwise and summing the voltages around the loop, we obtain

$$-v_1-10+v_2=0 \qquad (3.2.3)$$

Equations (3. 2. 1) and (3. 2. 3) form an independent set that can be used to solve for v_1 and v_2.

3.2.4 Circuits with controlled sources

Controlled sources present a slight additional complication of the node-voltage technique. In applying node-voltage analysis, first we write equations exactly as we have done for networks with independent sources. Then, we express the controlling variable in terms of the node-voltage variables and substitute into the network equations. We illustrate this with two examples.

Example 3.2.2 Node-voltage analysis with a dependent current source

Write an independent set of equations for the node voltages shown in Figure for example 3.2.2.

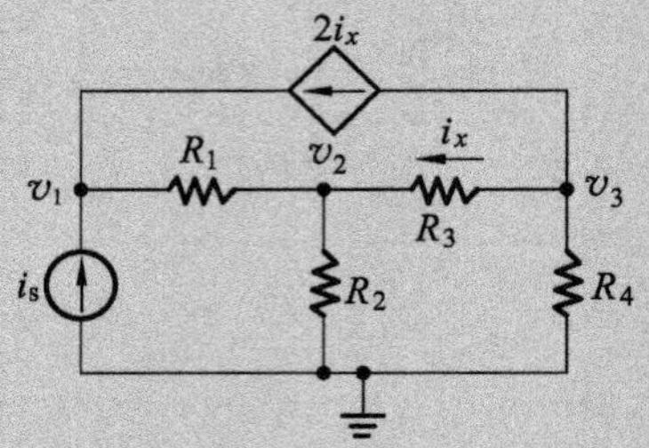

Figure for example 3.2.2 Circuit containing a current-controlled current source

Solution: First, we write KCL equations at each node, including the current of the controlled source just as if it were an ordinary current source:

$$(v_1-v_2)/R_1=i_s+2i_s$$
$$(v_2-v_1)/R_1+v_2/R_2+(v_2-v_3)/R_3=0$$
$$(v_3-v_2)/R_3+v_3/R_4+2i_s=0$$

Next, we find an expression for the controlling variable i_x in terms of the node voltages. Notice that i_x is the current flowing away from node 3 through R_3. Thus, we can write

$$i_x=(v_3-v_2)/R_3$$

Finally, we obtain the required equation set, 4 unknown quantities and 4 equations, and the problem is resolved.

Example 3.2.3 Node-voltage analysis with a dependent voltage source

Write an independent set of equations for the node voltages shown in Figure for example 3.2.3.

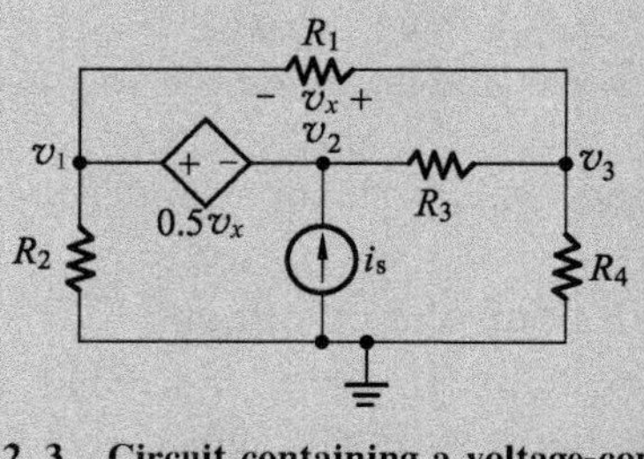

Figure for example 3.2.3 Circuit containing a voltage-controlled voltage source

Solution: First, we use KCL to write independent current equations. For a Supernode enclosing the controlled voltage source,

$$V_1/R_2+(v_1-v_3)/R_1+(v_2-v_3)/R_3=i_s$$

For node 3,

$$V_3/R_4+(v_3-v_2)/R_3+(v_3-v_1)/R_1=0$$

Now, among the 4 nodes, we already use 3 nodes. We do not need to write more current equations. Even we write equations, they are not independent.

Next, we use KVL to write voltage equations.

$$-v_1+0.5v_x+v_2=0$$

$$-v_1-v_x+v_3=0$$

Finally, 2 KCL equations and 2 KVL equations, we obtain the required equation set, 4 unknown quantities, 4 equations, and the problem is resolved.

Using the principles we have discussed in this section, we can write node-voltage equations for any network consisting of sources and resistances. We summarize the steps in analyzing circuits by the node-voltage technique:

(1) Select a reference node and assign variables for the unknown node voltages. If the reference node is chosen at one end of an independent voltage source, one node voltage is known at the start, and fewer need to be computed.

(2) Write network equations. First, use KCL to write current equations for nodes and Supernodes. Write as many currents as you can without using all of the nodes. Then if you do not have enough equations because of voltages sources connected between nodes, use KVL to write additional equations.

(3) If the circuit contains dependent sources, find expressions for the controlling variables in terms of the node voltages. Substitute expression into the network equations, and obtain equations having only the node voltages as unknowns.

(4) Put the equations into standard form and solve for the node voltages.

(5) Use the values found for the node voltages to calculate any other current or voltage of interest.

3.3 Mesh-current method

In this section, we will present how to analyze networks by using another general technique, known as **mesh-current analysis**.

3.3.1 Equations to solve for mesh currents

Let us start by considering the planar network shown in Figure 3.3.1(a). We need to find the currents. First, write equations for the currents shown in Figure 3.3.1 (a), which are called branch currents because a separate current is defined in each branch of the network. We will eventually see that using the mesh currents illustrated in Figure 3.3.1(b) makes the solution easier.

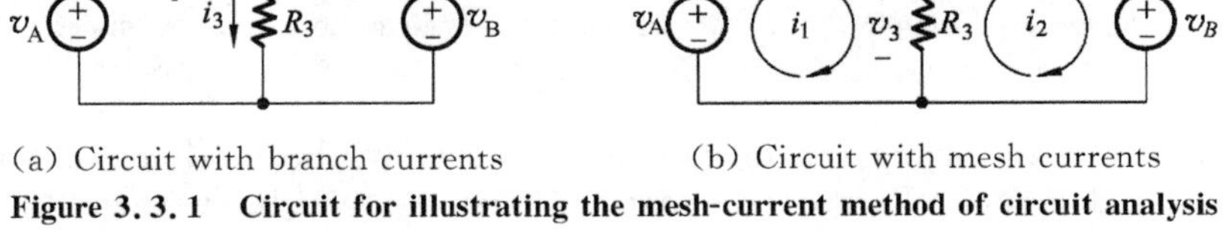

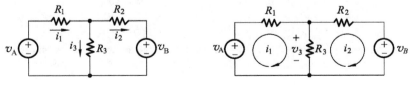

(a) Circuit with branch currents (b) Circuit with mesh currents

Figure 3.3.1 Circuit for illustrating the mesh-current method of circuit analysis

Three independent equations are needed to solve for the three branch currents shown in Figure 3. 3. 1(a). In general, the number of independent KVL equations that can be written for a planar network is equal to the number of meshes. If we have n nodes, b branches, we also can use $b-(n-1)$ to find the independent KVL equations.

Application of KVL to the mesh consisting of v_A, R_1, and R_3 yields

$$R_1 i_1 + R_3 i_3 = v_A \tag{3.3.1}$$

Similarly, for the mesh consisting of R_3, R_2, and v_B, we get

$$-R_3 i_3 + R_2 i_2 = -v_B \tag{3.3.2}$$

Applying KCL to the node at the top end of R_3, we have

$$i_1 = i_2 + i_3 \tag{3.3.3}$$

Substitute Equation (3. 3. 3) to Equations (3. 3. 1) and (3. 3. 2), we get

$$R_1 i_1 + R_3 (i_1 - i_2) = v_A \tag{3.3.4}$$

$$-R_3 (i_1 - i_2) + R_2 i_2 = -v_B \tag{3.3.5}$$

Now, consider the mesh currents i_1 and i_2 shown in Figure 3. 3. 1(b). As indicated in the figure, mesh currents are considered to flow around closed paths. Hence, mesh currents automatically satisfy KCL. When several mesh currents flow through one element, we consider the current in that element to be the algebraic sum of the mesh currents. Thus, assuming a reference direction pointing downward, the current in R_3 is $(i_1 - i_2)$. Thus, $v_3 = R_3(i_1 - i_2)$. Now if we follow i_1 around its loop and apply KVL, we get Equation (3. 3. 4) directly. Similarly, we obtain Equation (3. 3. 5) directly.

Because mesh currents automatically satisfy KCL, some work is saved in writing and solving the network equations. in more complex networks, the advantage can be quite significant.

Example 3.3.1 Mesh-Current Analysis

Write the equations needed to solve the mesh current in Figure for example 3.3.1(a).

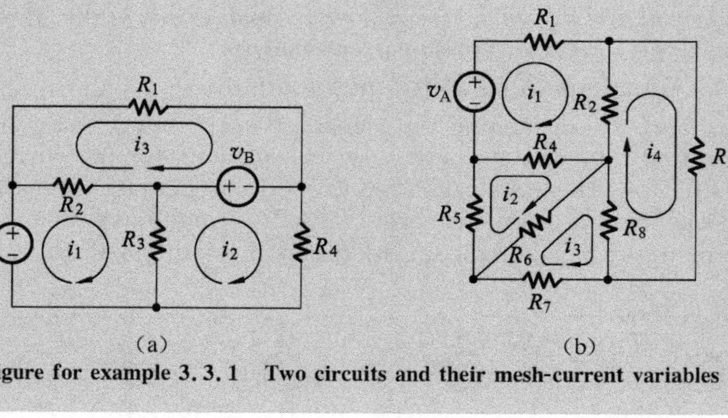

Figure for example 3. 3. 1 Two circuits and their mesh-current variables

Solution: Using a pattern in solving network by the mesh-current method helps to avoid errors. Part of the pattern that we use is to select the mesh currents to flow clockwise. Then, we write a KVL equation for each mesh, going around the meshes clockwise. As usual, we add a voltage if its positive reference is encountered first in traveling around the mesh, and we subtract the voltage if the negative reference is encountered first. Our pattern is always to take the first end of each resistor encountered as the positive reference for its voltage. Thus, we are always adding the resistor voltages.

For example, in mesh 1 of Figure for example 3.3.1(a), we first encounter the left-hand end of R_2. The voltage across R_2 referenced positive on its left-hand end is $R_2(i_1-i_3)$. Similarly, we encounter the top end of R_3 first, and the voltage across R_3 referenced positive at the top end is $R_3(i_1-i_2)$. By using the pattern, we add a term for each resistor in the KVL equation, consisting of the resistance times the current in the mesh under consideration minus the current in the adjacent mesh. Using this pattern for mesh 1, we have

$$R_2(i_1-i_3)+R_3(i_1-i_2)-v_A=0$$

Similarly, for mesh 2, we obtain

$$R_3(i_2-i_1)+R_4 i_2+v_B=0$$

Finally, for mesh 3, we have

$$R_2(i_3-i_1)+R_1 i_3-v_B=0$$

Notice that we have taken the positive reference for the voltage across R_3 at the top in writing the equation for mesh 1 and at the bottom for mesh 3. This is not an error because the terms for R_3 in the two equations are opposite in sign.

In standard form, the equations become:

$$(R_2+R_3)i_1-R_3 i_2-R_2 i_3=v_A$$
$$-R_3 i_1+(R_3+R_4)i_2=-v_B$$
$$-R_2 i_1+(R_1+R_2)i_3=v_B$$

In matrix form, we have

$$\begin{bmatrix}(R_2+R_3) & -R_3 & -R_2\\ -R_3 & (R_3+R_4) & 0\\ -R_2 & 0 & (R_1+R_2)\end{bmatrix}\begin{bmatrix}i_1\\ i_2\\ i_3\end{bmatrix}=\begin{bmatrix}v_A\\ -v_B\\ v_B\end{bmatrix}$$

Through this matrix form, we can find the rule: in mesh 1's equation, the coefficient of i_1 is the sum of resistances of mesh 1; the coefficient of i_2 is the resistor both in mesh 1 and mesh 2, but the value is negative; same coefficient of i_3, R_2 also is in mesh 1 and mesh 3, and the value is also negative. Same rule to mesh 2 and mesh 3's mesh-current equation.

Often, we use $\boldsymbol{R}$ to represent the coefficient matrix, $\boldsymbol{I}$ to represent the column vector of mesh currents, and V to represent the column vector of the terms on the right-hand side of the equations in standard form. Then, the mesh-current equations are represented as:

$$\boldsymbol{RI}=V$$

We refer to the element of the i_{th} row and j_{th} column of R as r_{ij}.

3.3.2 Mesh equations using matrix form

If a circuit contains only resistances and independent voltage sources, and if we select the mesh currents flowing clockwise, the mesh equations can be obtained directly in matrix form using these steps:

(1) Make sure that the circuit contains only resistances and independent voltage sources. Select all of the mesh currents to flow in the clockwise direction.

(2) Write the sum of the resistances contained in each mesh as the corresponding element on the main diagonal of **R**. In other words, r_{jj} equals the sum of the resistances encountered in going around mesh j.

(3) Insert the negative of the resistances common to the corresponding meshes as the off diagonal terms of **R**. Thus, for $i=1/j$, the elements r_{ij} and r_{ji} are the same and equal to negative of the sum of the resistances common to meshes i and j.

(4) For each element of the **V** matrix, go around the corresponding mesh clockwise, subtracting the values of voltage sources for which we encounter the positive reference first and adding the values of voltage sources for which we encounter the negative reference first. (We have reversed the rules for adding or subtracting the voltage source values from what we used when writing KVL equations, because the elements of V correspond to terms on the opposite side of the KVL equations.)

Keep in mind that this procedure does not apply to circuits having current sources or controlled sources.

Example 3.3.2

Write the equations for the mesh currents in Figure for example 3.3.1(b).

Use the same rule in Example 3.3.1, we can get the matrix form directly.

$$\begin{bmatrix} (R_1+R_2+R_4) & -R_4 & 0 & -R_2 \\ -R_4 & (R_4+R_5+R_6) & -R_6 & 0 \\ 0 & -R_6 & (R_6+R_7+R_8) & -R_8 \\ -R_2 & 0 & -R_8 & (R_2+R_3+R_8) \end{bmatrix} \begin{bmatrix} i_1 \\ i_2 \\ i_3 \\ i_4 \end{bmatrix} = \begin{bmatrix} v_A \\ 0 \\ 0 \\ 0 \end{bmatrix}$$

3.3.3 Circuits with current sources

When a circuit contains a current source, we must depart from the pattern that we use for circuits consisting of voltage sources and resistances. First, consider the circuit of Figure 3.3.2. As usual, we have defined the mesh currents flowing clockwise. If we were to try to write a KVL equation for mesh 1, we would need to include an unknown for the voltage across the current sources. In the circuit in Figure 3.3.2, we have defined the current in the current source as i_1. However, we know that this current is 2 A. Thus, we can write

$$i_1=2\ \text{A} \tag{3.3.6}$$

The second equation needed can be obtained by applying KVL to mesh 2, which yields

$$10(i_2-i_1)+5i_2+10=0 \tag{3.3.7}$$

Equations (3.3.6) and (3.3.7) can readily be solved for i_2. Notice that in this case the presence of a current source facilitates the solution.

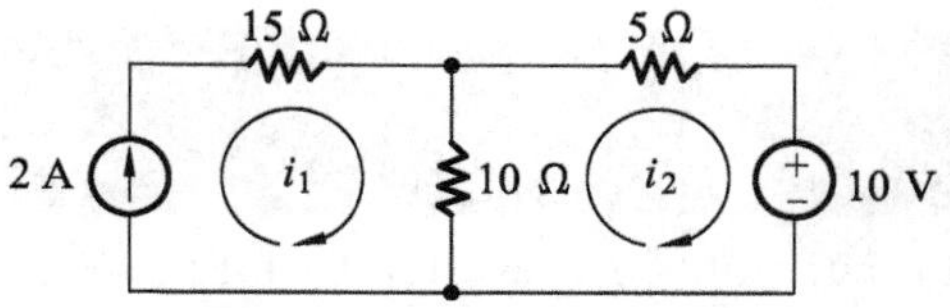

Figure 3. 3. 2 In this circuit, we have $i_1=2$ A

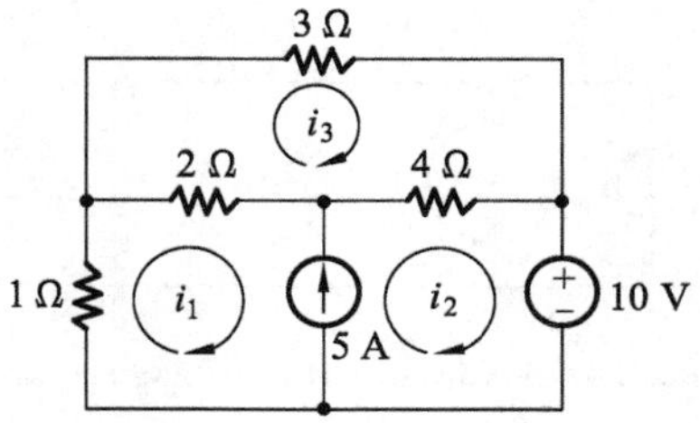

Figure 3. 3. 3 A circuit with a current source common to two meshes

Now let us consider the somewhat more complex situation shown in Figure 3. 3. 3. As usual, we have defined the mesh currents flowing clockwise. We cannot write KVL equation around mesh 1 because the voltage across the 5 A current source is unknown (and we do not want to increase the number of unknown in our equations). A solution is to combine meshes 1 and 2 into a **Supermesh.** In other words, we write a KVL equation around the periphery of meshs 1, 2 and combined. This yields

$$i_1+2(i_1-i_3)+4(i_2-i_3)+10=0 \tag{3. 3. 8}$$

Next, we can write a KVL equation for mesh 3:

$$3i_3+4(i_3-i_2)+2(i_3-i_1)=0 \tag{3. 3. 9}$$

Finally, we recognize that we have defined the current in the current source referenced upward as i_2-i_1. However, we know that the current flowing upward through the current source is 5 A. Thus, we have

$$i_2-i_1=5\text{ A} \tag{3. 3. 10}$$

It is important to realize that Equation (3. 3. 10) is not a KCL equation. Instead, it simply states that we have defined the current referenced upward through the current source in terms of the mesh currents as i_2-i_1, but this current is known to be 5 A. Equations (3. 3. 8), (3. 3. 9), and (3. 3. 10) can be solved for the mesh currents.

3.3.4 Circuits with controlled sources

Controlled sources present a slight additional complication to the mesh-current technique. First, we write equations exactly as we have done for networks with independent sources. Then, we express the controlling variables in terms of the mesh-current variables and substitute into the network equations. We illustrate this with an example.

Example 3. 3. 3　Mesh-current analysis with controlled sources

Solve for the currents in the circuit of Figure for example 3.3.3, which contains a voltage-controlled current source common to the two meshes.

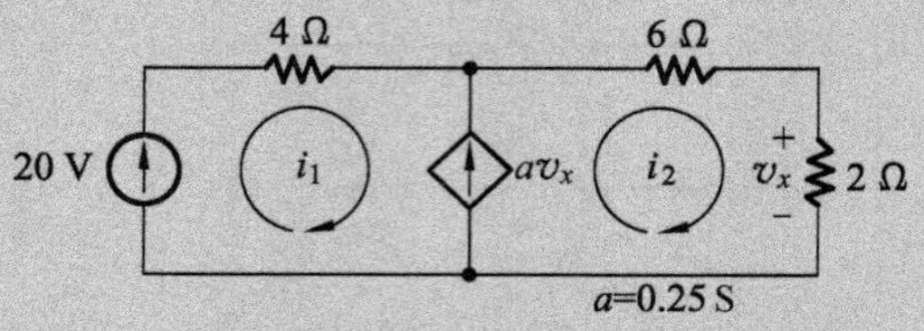

Figure for example 3. 3. 3　Circuit with a voltage-controlled current source

Solution: First, we write equations for the mesh currents as we have done for independent sources, since there is a current source common to mesh 1 and mesh 2. We start by combining the meshes to form a Supermesh and write a voltage equation:

$$-20+4i_1+6i_2+2i_2=0$$

Then, we write an expression for the source current in terms of the mesh currents:

$$av_x=0.25v_x=i_2-i_1$$

Next, we see that the controlling voltage $v_x=2i_2$. Using this equation set, we can solve the equations which yield $i_1=1$ A and $i_2=2$ A.

Using the principle we have discussed in this section, we can write mesh-current equations for any planar network consisting of sources and resistances.

We summarize the steps in analyzing planar circuits by the mesh-current technique:

(1) If necessary, redraw the network without crossing conductors or elements. Then, define the mesh current flowing around each of the mesh defined by the network. For consistency, we usually select a clockwise direction for each of the mesh currents, but this is not a requirement.

(2) Write network equations. First, use KVL to write voltage equations for meshes that do not contain current source. Next, if any current source is present, write expressions for their current in terms of the mesh currents. Finally, if a current source is common to two meshes, write a KVL equation for the Supermesh.

(3) If the circuit contains dependent sources, find expression for the controlling variables in terms of the mesh currents. Substitute into the network equations, and obtain equations having only the mesh currents as unknowns.

(4) Put the equations into standard form. Solve for the mesh currents by use of determinations or other means.

(5) Use the values for the mesh currents to calculate any other current or voltage of interest.

3.4 Node-voltage method versus Mesh-current method

The greatest advantage of both the node-voltage and mesh-current methods is that they reduce the number of simultaneous equations that must be manipulated. They also require the analyst to be quite systematic in terms of organizing and writing these equations. It is natural to ask, then, "when is the node-voltage method preferred to the mesh-current method and vice versa?" As you might suspect, there is no clear-cut answer. Asking a number of questions, however, may help you to identify the more efficient method before plunging into the solution process:

Does one of the methods result in fewer simultaneous equations to solve?

Does the circuit contain Supernodes? If so, using the node-voltage method will permit you to reduce the number of equations to be solved.

Does the circuit contain Supermeshes? If so, using the mesh-current method will permit you to reduce the number of equations to be solved.

Will solving some portion of the circuit give the requested solution? If so, which method is the most efficient for solving just the pertinent portion of the circuit?

Perhaps the most important observation is that, for any situation, some time spent thinking about the problem in relation to the various analytical approaches available is time well spent. Examples 3. 4. 1 and 3. 4. 2 illustrate the process of deciding between the node-voltage and mesh-current methods.

Example 3. 4. 1 Understanding the node-voltage method versus mesh-current method

Find the power dissipated in the 300 Ω resistor in the circuit shown in Figure A for example 3.4.1.

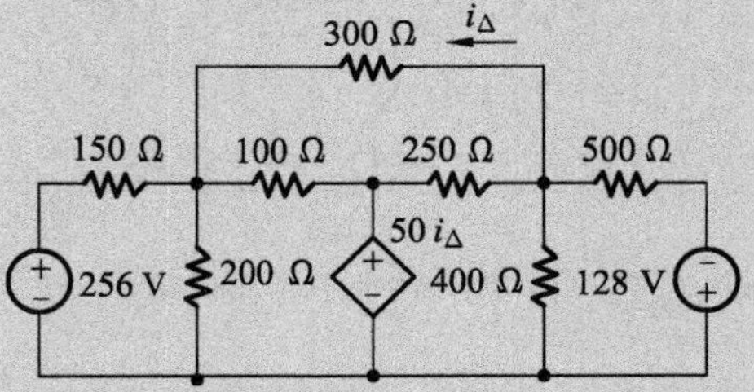

Figure A for example 3. 4. 1

Solution: To find the power dissipated in the 300 Ω resistor, we need to find either the current in the resistor or the voltage across it.

The mesh-current method yields the current in resistor; this approach requires solving five simultaneous mesh equations, as depicted in Figure B for example 3. 4. 1. In writing the five equations, we must include the constraint $i_\Delta = -i_b$.

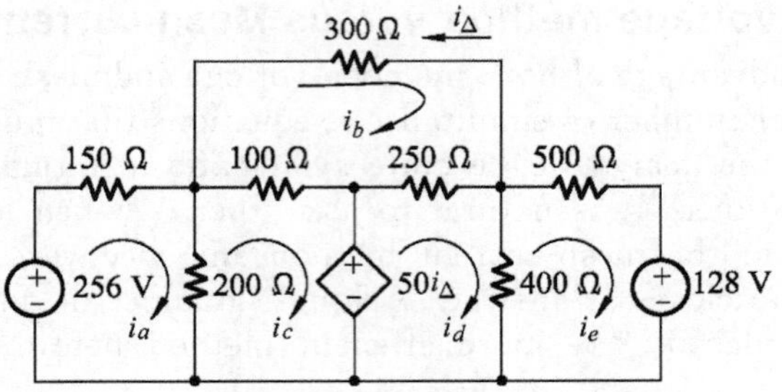

Figure B for example 3. 4. 1

Before going further, let's also look at the circuit in terms of the node-voltage method. Note that, once we know the node-voltages, we can calculate either the current in the 300 Ω resistor or the voltage across it. The circuit has 4 essential nodes, and therefore only 3 node-voltage equations are required to describe the circuit. Because of the dependent voltage source between two essential nodes, we have to sum the currents at only two nodes. Hence the problem is reduced to that writing 2 node-voltage method requires only 3 simultaneous equations, which is a more attractive approach.

Once the decision to use the node-voltage method has been made, the next step is to select a reference node. Then label the other nodes, shown in Figure C for example 3. 4. 1. We defined the three node voltages v_1, v_2, and v_3, which also indicated that nodes 1 and 3 form a Supernode, because they are connected by a dependent voltage source.

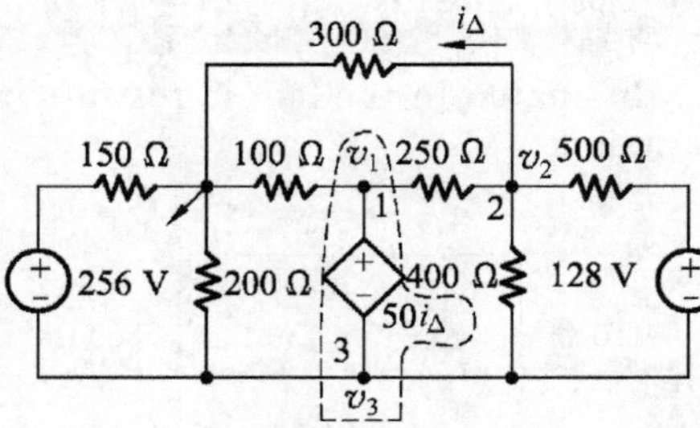

Figure C for example 3. 4. 1 The circuit with a reference node (change the symbol of ground)

At the supernode,

$$\frac{v_1}{100}+\frac{v_1-v_2}{250}+\frac{v_3}{200}+\frac{v_3-v_2}{400}+\frac{v_3-(v_2+128)}{500}+\frac{v_3+256}{150}=0$$

At v_2,

$$\frac{v_2}{300}+\frac{v_2-v_1}{250}+\frac{v_2-v_3}{400}+\frac{v_2+128-v_3}{500}=0$$

From the Supernode, the constraint equation is

$$v_3=v_1-50i_\Delta=v_1-\frac{v_2}{6}$$

The second node that merits consideration as the reference node is the lower node in the circuit, as shown in Figure D for example 3. 4. 1.

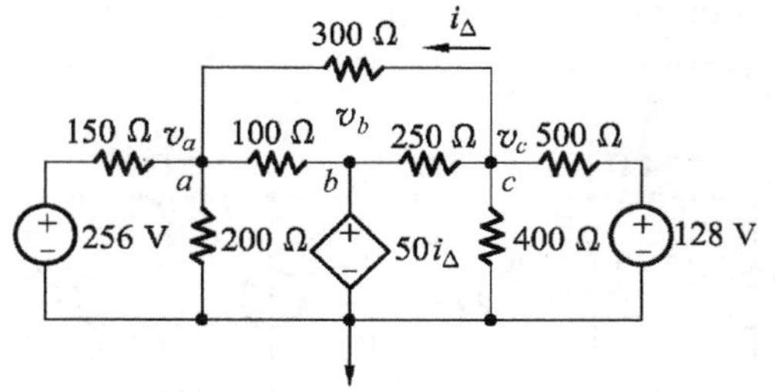

Figure D for example 3.4.1 The circuit with an alternative reference node (change the symbol of ground)

At v_a,

$$\frac{v_a}{200}+\frac{v_a-256}{150}+\frac{v_a-v_b}{100}+\frac{v_a-v_c}{300}=0$$

At v_c,

$$\frac{v_c}{400}+\frac{v_c+128}{500}+\frac{v_c-v_b}{250}+\frac{v_c-v_a}{300}=0$$

From the Supernode, the constraint equation is

$$v_b=50i_\Delta=50\frac{v_c-v_a}{300}=\frac{v_c-v_a}{6}$$

You should verify that the solution of either set leads to a power calculation of 16.57 W dissipated in the 300 Ω resistor.

Example 3.4.2 Comparing the node-voltage and mesh-current methods

Find the voltage v_a in the circuit shown in Figure A for example 3.4.2.

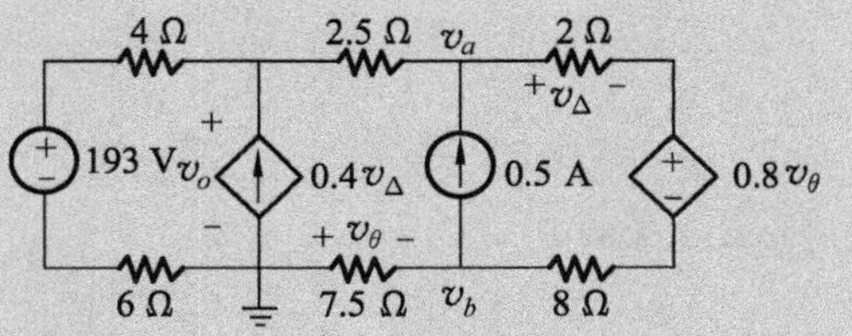

Figure A for example 3.4.2 The circuit

Solution: At the first glance, the node-voltage method looks appealing, because we may define the unknown voltage as a node-voltage by choosing the lower terminal of the dependent current source as the reference node. The circuit has 4 essential nodes and 2 voltage-controlled dependent sources, so the node-voltage method requires manipulation of 3 node-voltage equations and 2 constraint equations.

$$193=10i_a+10i_b+10i_c+0.8v_\theta$$

$$i_b-i_a=0.4v_\Delta=0.8i_c$$

To help you to compare the two approaches, we summarize both methods. The mesh-current equations are based on the circuit shown in Figure B for example 3.4.2.

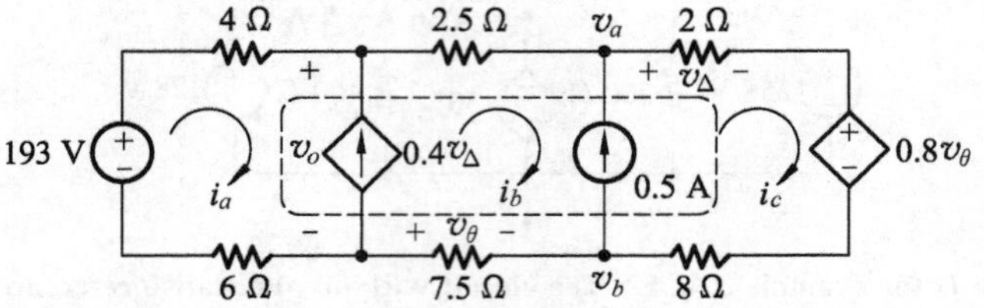

Figure B for example 3.4.2 The circuit with three mesh currents

And the constraint equations are

$$v_\theta = 0.75 i_b$$
$$i_c - i_b = 0.5$$

We use constraint equation to write the Supermesh equation in terms of i_a:

$$160 = 80 i_a, \text{ or } i_a = 2 \text{ A}$$
$$v_o = 193 - 20 = 173 \text{ V}$$

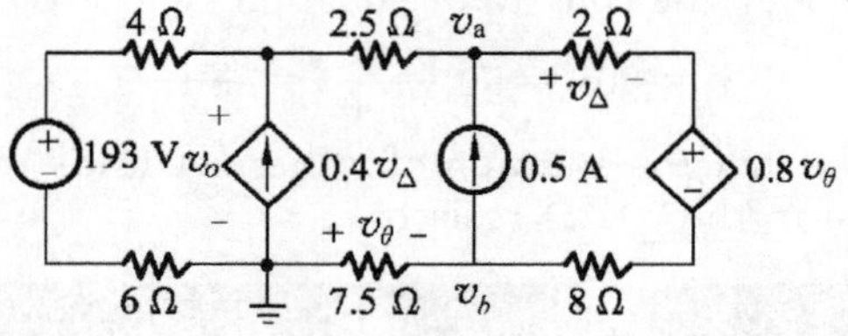

Figure C for example 3.4.2 The circuit with node voltage

The node-voltage equations are

$$\frac{v_o - 193}{10} - 0.4v_\Delta + \frac{v_o - v_a}{2.5} = 0$$
$$\frac{v_o - v_a}{2.5} - 0.5 + \frac{v_a - (v_b + 0.8v_\theta)}{10} = 0$$
$$\frac{v_b}{7.5} + 0.5 + \frac{v_b + 0.8v_\theta - v_a}{10} = 0$$

The constraint equations are

$$v_\theta = -v_b, \quad v_\Delta = 2\frac{v_a - (v_b + 0.8v_\theta)}{10}$$

The node-voltage approach also gives

$$v_o = 173 \text{ V}$$

3.5 Source transformation

We can replace a voltage source in series with a resistor by a current source in parallel with a resistor. This is called **source transformation** and is illustrated in Figure 3.5.1.

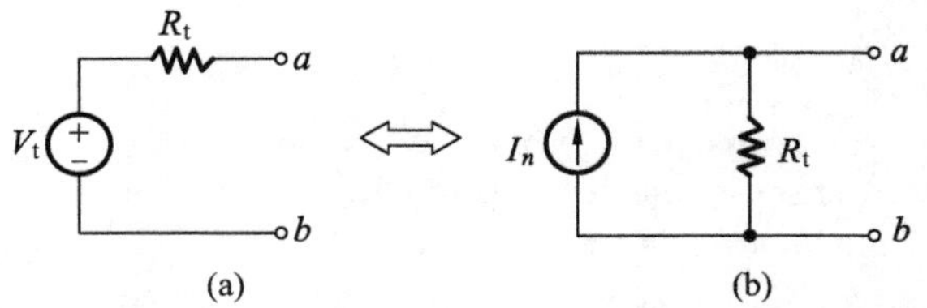

Figure 3. 5. 1 A voltage source in series with a resistor is externally equivalent to a current source in parallel with a resistor

The double headed arrow emphasizes that a source transformation is bilateral; that is, we can start with either configuration and derive the other.

We need to find the relationship between v_s and i_s that guarantees the two configurations in Figure 3. 5. 1 are equivalent with respect to nodes a, b. Equivalence is achieved if any resistor R_L experiences the same current flow, and thus the same voltage drop, whether connected between nodes a, b in Figure 3. 5. 1 (a) or (b).

Suppose R_L is connected between nodes a, b in Figure 3. 5. 1(a). Using Ohm's law, the current in R_L is

$$i_L = \frac{v_s}{R_t + R_L} \tag{3. 5. 1}$$

Now suppose the same resistor R_L is connected between nodes a, b in Figure 3. 5. 1(b). Using current division, the current in R_L is

$$i_L = \frac{R_t}{R_t + R_L} i_s \tag{3. 5. 2}$$

If the two circuits in Figure 3. 5. 1 are equivalent, these resistor currents must be the same. Equate the right-hand sides of Equations (3. 5. 1) and (3. 5. 2) and simplify them.

$$i_s = \frac{v_s}{R} \tag{3. 5. 3}$$

When Equation (3. 5. 3) is satisfied for the circuits in Figure 3. 5. 1, the current in R_L is the same for both circuits in the figure for all values of R_L. If the current through R_L is the same in both circuits, then the voltage drop across R_L is the same in both circuits, and the circuits are equivalent at nodes a, b.

In making source transformations it is very important to maintain the proper relationship between the reference direction for the current source and polarity of the voltage source. If the positive polarity is closest to terminal a, the current reference must point toward terminal a, as shown in Figure 3. 5. 1.

Sometimes, we can simplify the solution of a circuit by source transformations. This is similar to solving circuits by combining resistors in series or parallel.

Example 3.5.1

Using source transformations

Use source transformations to aid in solving for the currents i_1 and i_2 shown in Figure for example 3.5.1(a).

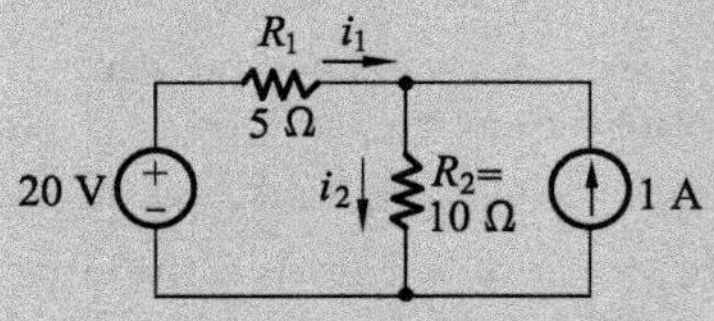

(a) Original circuit

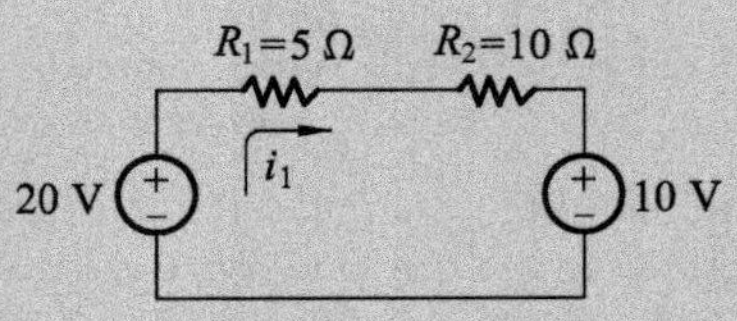

(b) Circuit after transforming the current source into a voltage source

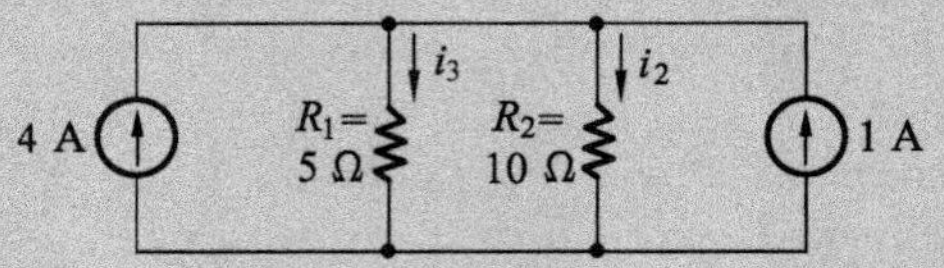

(c) Circuit after transforming the voltage soure into a current soure

Figure for example 3. 5. 1

Solution: Several approaches are possible. One is to transform the 1 A current source and R_2 into a voltage source in series with R_2. This is shown in Figure for example 3. 5. 1(b). Notice that the positive polarity of the 10 V source is at the top, because the 1 A source reference points upward. The single-loop circuit of Figure for example 3. 5. 1(b) can be solved by writing a KVL equation. Traveling clockwise and summing voltages, we have

$$R_1 i_1 + R_2 i_1 + 10 - 20 = 0$$

Solving and substituting values, we get

$$i_1 = \frac{10}{R_1 + R_2} = 0.667\ \text{A}$$

$$i_2 = i_1 + 1 = 1.667\ \text{A}$$

Another approach is to transform the voltage source and R_1 into a current source in parallel with R_1. Making this change to the original circuit yields the circuit shown in Figure for example 3. 5. 1(c). Notice that we have labeled the current through R_1 as i_3 rather than i_1. This is because the current in resistor of the transformed source is not the same as in the original circuit. Now, in Figure for example 3. 5. 1(c), we see that a total current of 5 A flow into the parallel combination of R_1 and R_2. Using the current-division principle, we find the current through R_2:

$$i_2 = \frac{R_1}{R_1 + R_2} i_{\text{total}} = 1.667\ \text{A}$$

This agrees with our previous result.

Example 3.5.2

Using source transformations to solve a circuit

(a) For the circuit shown in Figure A for example 3.5.2, find the power associated with the 6 V source.

(b) State whether the 6 V source is absorbing or delivering the power.

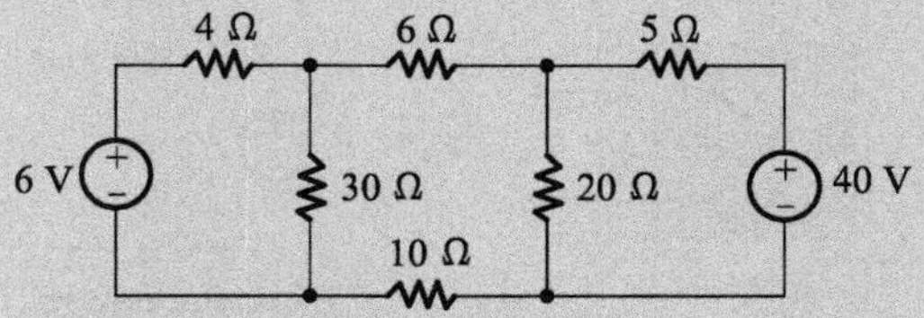

Figure A for example 3.5.2

Solution: (a) We reduce the circuit in a way that we preserve the identity of the branch containing the 6 V source. Then we reduce the circuit step by step shown in Figure B for example 3.5.2.

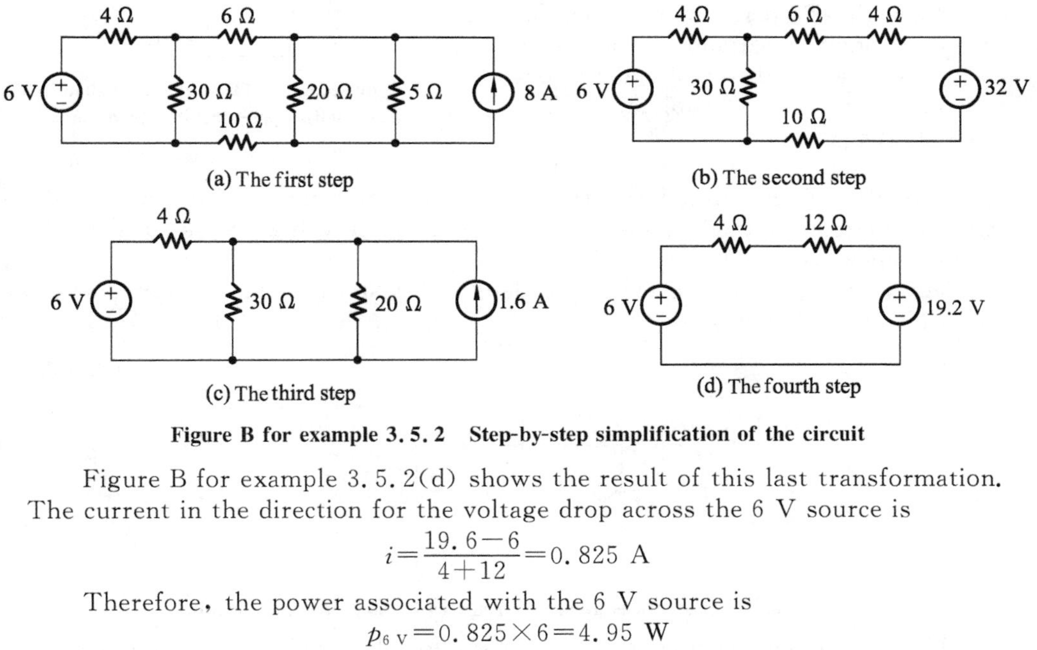

Figure B for example 3.5.2 Step-by-step simplification of the circuit

Figure B for example 3.5.2(d) shows the result of this last transformation. The current in the direction for the voltage drop across the 6 V source is

$$i=\frac{19.6-6}{4+12}=0.825\ \text{A}$$

Therefore, the power associated with the 6 V source is

$$p_{6\ \text{V}}=0.825\times 6=4.95\ \text{W}$$

(b) The direction of current flows into the 6 V source is the same as the direction of voltage drop of 6 V source, so the voltage source is absorbing power.

3.6 Thevenin and Norton theorem

In this section, we learn how to replace two-terminal circuits containing resistances and sources by simple equivalent circuits. By a two-terminal circuit, we mean that the original circuit has only two points that can be connected to other circuits. The original circuit can be any complex interconnection of resistances and sources. However, a restriction is that the controlling variables for any controlled sources must appear inside the original circuit.

3.6.1 Thevenin equivalent circuit

Thevenin equivalent circuit is a circuit consisting of an independent voltage source in series with a resistance. This is illustrated in Figure 3.6.1.

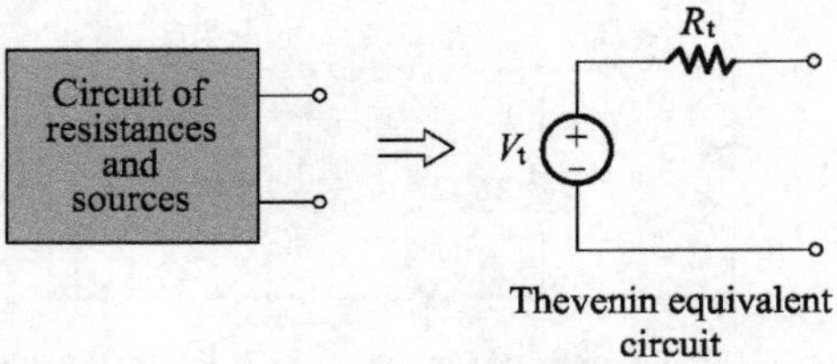

Figure 3.6.1 Two-terminal circuit consisting of resistances and sources can be replaced by a Thevenin equivalent circuit

Consider the Thevenin equivalent circuit with open-circuited terminals as shown in Figure 3.6.2. By definition, no current can flow through an open circuit. Therefore, no current flows through the Thevenin resistance, and the voltage across the resistance is zero. Applying KVL, we conclude that

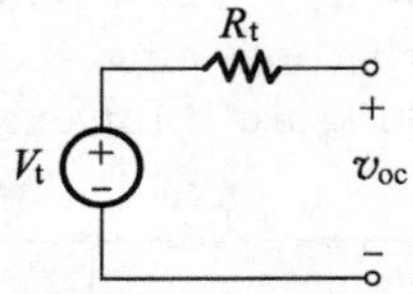

Figure 3.6.2 Thevenin equivalent circuit with open-circuited terminals

$$V_t = v_{oc} \tag{3.6.1}$$

Both the original circuit and the equivalent circuit are required to have the same open-circuit voltage. Thus, the Thevenin source voltage V_t is equal to the open-circuit voltage of the original network.

Now, consider the Thevenin equivalent with a short circuit connected across its terminals as shown in Figure 3.6.3. The current flowing in this circuit is

$$i_{sc} = \frac{V_t}{R_t} \tag{3.6.2}$$

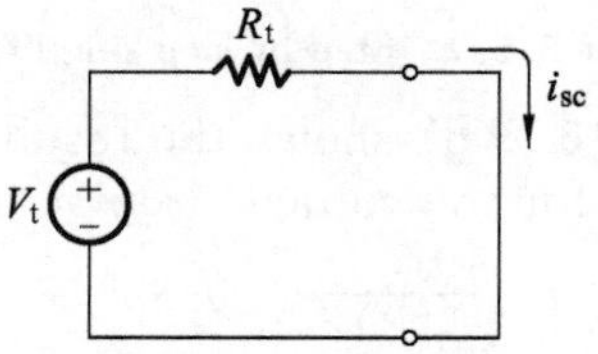

Figure 3.6.3 Thevenin equivalent circuit with short-circuited terminals

The short-circuit current i_{sc} is the same for the original circuit as for the Thevenin equivalent. Solving for the Thevenin resistance, we have

$$R_t = \frac{V_t}{i_{sc}} \tag{3.6.3}$$

Using the fact that the Thevenin voltage is equal to the open-circuit voltage of the network, we have

$$R_t = \frac{V_{oc}}{i_{sc}} \tag{3.6.4}$$

Thus, to determine the Thevenin equivalent circuit, we can start by analyzing the original network for its open-circuit voltage and short-circuit current.

Example 3.6.1 Finding the Thevenin equivalent circuit

Find the Thevenin equivalent for the circuit shown in Figure for example 3.6.1(a).

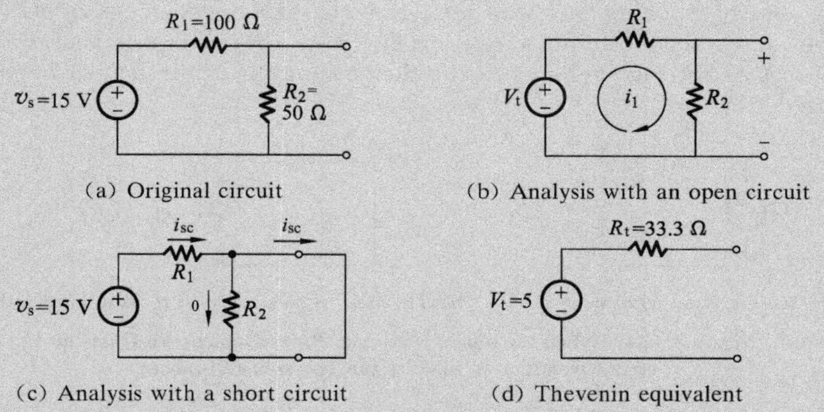

Figure for example 3.6.1

Solution: First, we analyze the circuit with open-circuited terminals. This is shown in Figure for example 3.6.1(b). The resistors R_1 and R_2 are in series and have an equivalent resistance of R_1+R_2. Therefore, the current calculating is

$$i_1 = \frac{v_s}{R_1+R_2} = 0.10\ \text{A}$$

The open-circuit voltage is the voltage across R_2:

$$v_{oc} = R_2 i_1 = 5\ \text{V}$$

Thus, the Thevenin voltage is $V_t = 5$ V.

Now, consider the circuit with a short circuit connected across its terminals, which is shown in Figure 3.6.4(c). By definition, the voltage across a short circuit is zero. Hence, the voltage across R_2 is zero, and the current through it is zero, as shown in the Figure for example 3.6.1. Therefore, the short-circuit current i_{sc} flows through. So we can write

$$i_{sc} = \frac{v_s}{R_1} = 0.15\ \text{A}$$

Now, we can use Equation (3.6.4) to determine the Thevenin resistance:

$$R_t = \frac{v_{oc}}{i_{sc}} = 33.3\ \Omega$$

The Therenin equiralent circuit is shown in Figure for example 3.6.1 (d).

3.6.2 Thevenin resistance

If a network contains no dependent sources, there is an alternative way to find the Thevenin resistance. First, we zero the sources in the network. In zeroing a voltage source, we reduce its voltage to zero. A voltage source with zero voltage is equivalent to a short circuit.

In zeroing a current source, we reduce its current to zero. By definition, an element that always carries zero current is an open circuit. Thus, to zero the independent sources, we replace voltage sources with short circuits and replace current sources with open circuits.

Figure 3.6.4 shows a Thevenin equivalent before and after zeroing its voltage source. Looking back into the terminals after the source is zeroed, we see the Thevenin resistance. Thus, we can find the Thevenin resistance by zeroing the sources in the original network and then computing the resistance between the terminals.

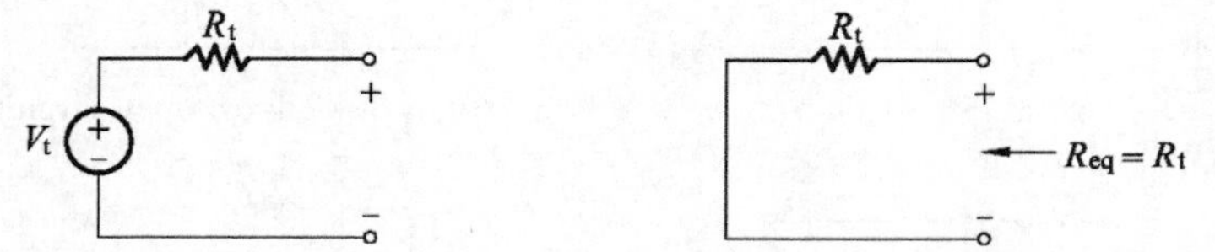

(a) Thevenin equivalent (b) Thevenin equivalent with its source zeroed

Figure 3.6.4 When the source is zeroed, the resistance seen from the circuit terminals is equal to the Thevenin resistance

Example 3.6.2 Zeroing sources to find Thevenin resistance

Find the Thevenin resistance for the circuit shown in Figure for example 3.6.2(a) by zeroing the sources. Then, find the short-circuit current and the Thevenin equivalent circuit.

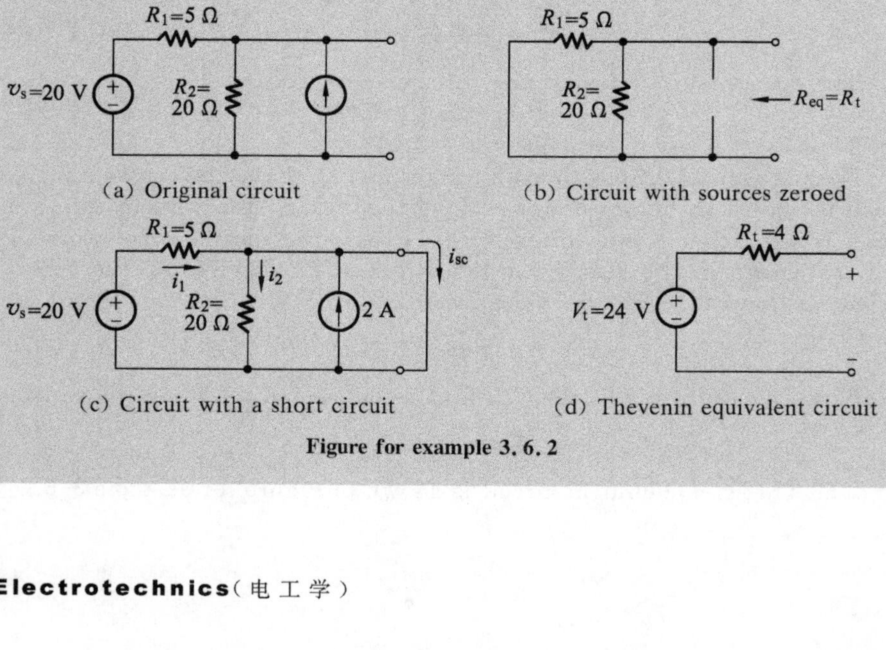

Figure for example 3.6.2

Solution: To zero the sources, we replace the voltage source by a short circuit and replace the current source by an open circuit. The resulting circuit is shown in Figure for example 3. 6. 2(b).

The Thevenin resistance is the equivalent resistance between the terminals. This is the parallel combination of R_1 and R_2, which is given by

$$R_t = R_{eq} = \frac{1}{1/R_1 + 1/R_2} = 4\ \Omega$$

Next, we find the short-circuit current for the circuit. The circuit is shown in Figure for example 3. 6. 2(c). In this circuit, the voltage across R_2 is zero because of the short circuit. Thus, the current through R_2 is zero:

$$i_2 = 0$$

Furthermore, the voltage across R_1 is equal to 20 V. Thus, the current is

$$i_1 = \frac{v_s}{R_1} = 4\ \text{A}$$

Finally, we write a current equation for the node joining the top ends of R_2 and the 2 A source. Setting the sum of the currents entering equal to the sum of the current leaving, we have

$$i_1 + 2 = i_2 + i_{sc}$$

This yields

$$i_{sc} = 6\ \text{A}$$

Now, the Thevenin voltage can be found. From Equation (3. 6. 3) to determine the Thevenin resistance:

$$V_t = R_t i_{sc} = 24\ \text{V}$$

The Thevenin equivalent circuit is shown in Figure for example 3. 6. 2(d).

Example 3. 6. 3 Thevenin equivalent of a circuit a dependent source

Find the Thevenin equivalent for the circuit shown in Figure for example 3.6.3(a).

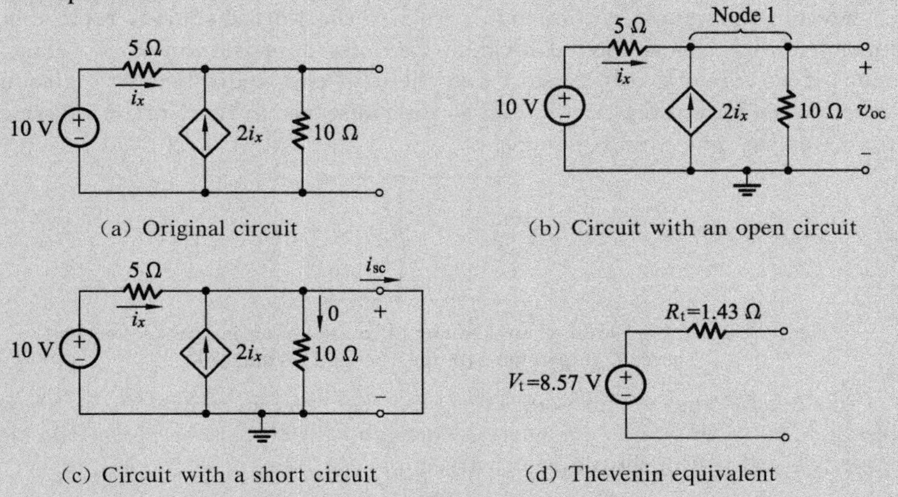

(a) Original circuit

(b) Circuit with an open circuit

(c) Circuit with a short circuit

(d) Thevenin equivalent

Figure for example 3. 6. 3

Solution: Because this circuit contains a dependent source, we cannot find the Thevenin resistance by zeroing the sources and combining resistors in series and parallel. Thus, we must analyze the circuit to find the open-circuit voltage and the short-circuit current.

We start with the open-circuit voltage. Consider Figure for example 3. 6. 3 (b). We use node voltage analysis, picking the reference node at the bottom of the circuit. Then, v_{oc} is the unknown node-voltage variable. First, we write a current equation at node 1.

$$i_x + 2i_x = \frac{v_{oc}}{10}$$

Next, we write an expression for the controlling variable i_x in terms of the node voltage v_{oc}:

$$i_x = \frac{10 - v_{oc}}{5}$$

Through these two equations, we have

$$v_{oc} = 8.57\ \text{V}$$

Now, we consider short-circuit conditions as shown in Figure for example 3. 6. 3(c). In this case, the current through the 10 Ω resistor is zero. Furthermore, we get

$$i_x = 2\ \text{A and } i_{sc} = 3i_x = 6\ \text{A}$$

Next, we use Equation (3. 6. 3) to compute the Thevenin resistance:

$$R_t = \frac{V_{oc}}{i_{sc}}$$

Finally, the Thevenin equivalent circuit is shown in Figure for example 3. 6. 3(d).

3.6.3 Norton equivalent circuit

Another type of equivalent, known as the **Norton equivalent circuit**, is shown in Figure 3. 6. 5. It consists of an independent current source I_n in parallel with the Thevenin resistance. Notice that if we zero the Norton source, replacing it by an open circuit, the Norton equivalent becomes a resistance of R_t. This also happens if we zero the voltage source in the Thevenin equivalent by replacing the voltage source by a short circuit. This, the resistance in the Norton equivalent is the same as the Thevenin resistance.

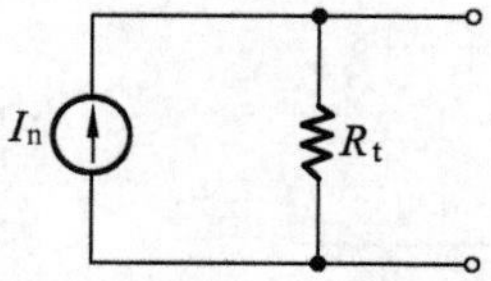

Figure 3. 6. 5 The Norton equivalent circuit consists of an independent current source I_n in parallel with the Thevenin resistance R_t

Consider placing a short circuit across the Norton equivalent as shown in Figure 3. 6. 6. In this case, the current through R_t is zero. Therefore, the Norton current is equal to the short-circuit current:

$$I_n = i_{sc}$$

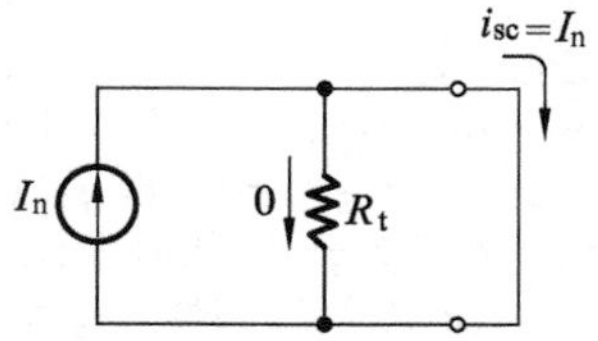

Figure 3.6.6 The Norton equivalent circuit with a short circuit across its terminals

We can find the Norton equivalent by using the same techniques as we used for the Thevenin equivalent.

3.6.4 Steps of Thevenin/Norton-equivalent analysis

(1) Perform two of these:

a. Determine the open circuit voltage $V_t=v_{oc}$.

b. Determine the short circuit current $I_n=i_{sc}$.

c. Zero the independent sources and find the Thevenin resistance R_t looking back into the terminals. Do not zero dependent sources.

(2) Use the equation $V_t=R_t I_n$ to compute the remaining value.

(3) The Thevenin equivalent consists of a voltage source V_t in series with R_t.

(4) The Norton equivalent consists of a current I_n in parallel with R_t.

3.7 Maximum power transfer

3.7.1 Maximum power transfer

Circuit analysis plays an important role in the analysis of systems designed to transfer power from a source to a load. We discuss power transfer in terms of two basic types of systems. The first emphasizes the efficiency of the power transfer. Power utility systems are a good example of this type because they are concerned with the generation, transmission, and distribution of large quantities of electric power. If a power utility system is inefficient, a large percentage of the power generated will be lost in the transmission and distribution processes, and thus wasted.

The second basic type of system emphasizes the amount of power transferred. Communication and instrumentation systems are good examples because in the transmission of information, or data, via electric signals, the power available at the transmitter or detector is limited. Thus, transmitting as much of this power as possible to the receiver or load, is desirable. In such applications the amount of power being transferred is small, so the efficiency of transfer is not a primary concern. We now consider maximum power transfer in systems that can be modeled by a pure resistive circuit.

Maximum power transfer can best be described with the aid of the circuit shown in Figure 3.7.1.

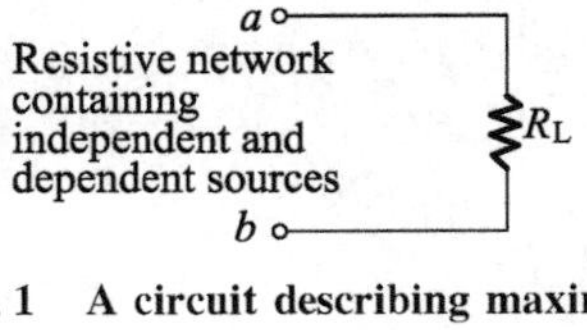

Figure 3.7.1 A circuit describing maximum power

We assume a resistive network containing independent and dependent sources and a designated pair of terminals a, b, to which a load R_L, is to be connected. The problem is to determine the value of R_L that permits maximum power delivery to R_L. The first step in this process is to recognize that a resistive network can always be replaced by its Thevenin equivalent. Therefore, we redraw the circuit shown in Figure 3.7.2. Replacing the original network by its Thevenin equivalent greatly simplifies the task of finding R_L. Derivation of R_L requires expressing the power dissipated in R_L as a function of the three circuit parameters V_{Th}, R_{Th}, and R_L.

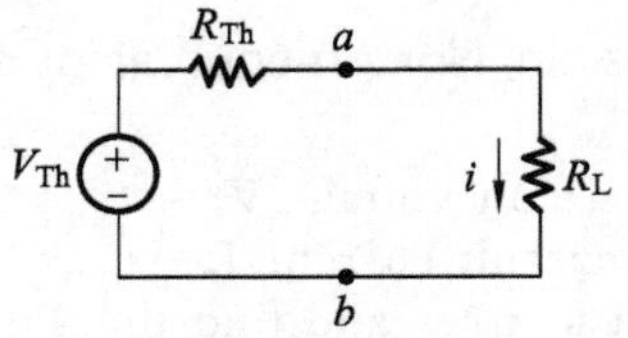

Figure 3.7.2 A circuit used to determine the value of R_L for maximum power transfer

Thus,

$$P=i^2R_L=\left(\frac{V_{Th}}{R_{Th}+R_L}\right)^2 R_L \tag{3.7.1}$$

Next, we recognize that for a given circuit, V_{Th} and R_{Th} will be fixed. Therefore, the power dissipated is a function of the single variable R_L. To find the value of R_L that maximizes the power, we use elementary calculus. We begin by writing an equation for the derivative of P with respect to R_L:

$$\frac{dp}{dR_L}=V_{Th}^2\left[\frac{(R_{Th}+R_L)^2-R_L\times 2(R_{Th}+R_L)}{(R_{Th}+R_L)^4}\right] \tag{3.7.2}$$

The derivative is zero and p is maximized when

$$(R_{Th}+R_L)^2=2R_L(R_{Th}+R_L) \tag{3.7.3}$$

Then we get

$$R_L=R_{Th} \tag{3.7.4}$$

Thus maximum power transfer occurs when the load resistance R_L equals the Thevenin resistance R_{Th}. To find the maximum power delivered to R_L, we simply substitute Equation (3.7.4) into the Equation (3.7.1):

$$P_{max}=\frac{V_{Th}^2R_L}{(2R_L)^2}=\frac{V_{Th}^2}{4R_L} \tag{3.7.5}$$

The analysis of a circuit when the load resistor is adjusted for maximum power transfer is illustrated in Example 3.7.1.

Example 3.7.1 Determining maximum power transfer

Find the load resistance for maximum power transfer from the circuit shown in Figure for example 3.7.1. Also, find the maximum power.

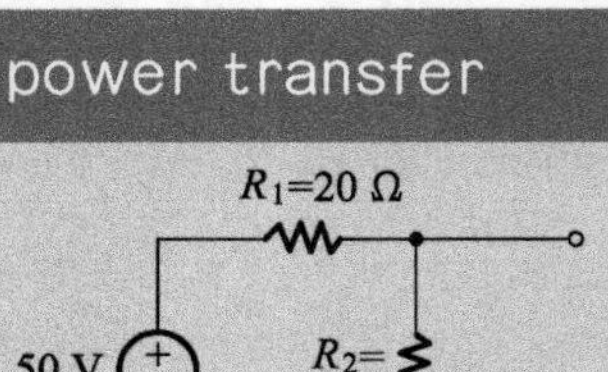

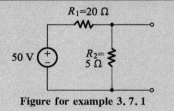

Figure for example 3.7.1

Solution: First, we must find the Thevenin equivalent circuit. Zeroing the voltage source, we find that the resistances R_1 and R_2 are in parallel. Thus, the Thevenin resistance is

$$R_t = \frac{1}{1/R_1 + 1/R_2} = 4\ \Omega$$

The Thevenin voltage is equal to the open-circuit voltage. Using the voltage-division principle, we find that

$$V_t = v_{oc} = \frac{R_2}{R_1 + R_2} \times 50 = 10\ \text{V}$$

Hence, the load resistance that receives maximum power is

$$R_L = R_t = 4\ \Omega$$

And the maximum power is given by Equation (3.7.5):

$$P_{L\max} = \frac{V_t^2}{4R_t} = 6.25\ \text{W}$$

3.7.2 Application of Maximum power transfer

When a load resistance equals the internal Thevenin resistance of the source, half of the power is dissipated in the source resistance and half is delivered to the load. In higher power applications for which efficiency is important, we do not usually design for maximum power transfer. For example, in designing an electric vehicle, we would want to deliver the energy stored in the batteries mainly to the derive motors and minimize the power loss in the resistance of the battery and wiring. This system would approach maximum power transfer rarely when maximum acceleration is needed.

On the other hand, when small amount of power is involved, we would design for maximum power transfer. For example, we would design a radio receiver to extract the maximum signal power from the receiving antenna. In this application, the power is very small, typically much less than one microwatt, and efficiency is not a consideration.

3.8 Superposition

A linear system obeys the principle of **superposition**, which states that whenever a linear system is excited, or driven, by more than one independent source of energy, the total response is the sum of the individual responses. An individual response is the result of an independent source acting alone.

We illustrate the validity of superposition with the example circuit shown in Figure 3.8.1.

In this circuit, there are two independent sources: the first, the voltage source v_{s1}, and the second, the current source i_{s2}. Suppose that the response of interest is the voltage across the resistance R_2.

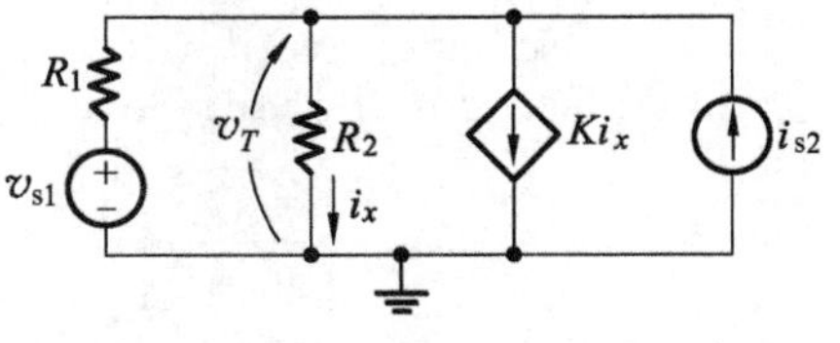

Figure 3.8.1 Circuit used to illustrate the superposition principle

First, we solve for the total response v_T by solving the circuit with both sources in place. Writing a current equation at the top node, we obtain

$$\frac{v_T - v_{s1}}{R_1} + \frac{v_T}{R_2} + Ki_x = i_{s2} \qquad (3.8.1)$$

The control variable i_x is given by

$$i_x = \frac{v_T}{R_2} \tag{3.8.2}$$

Substituting Equation (3. 8. 2) into Equation (3. 8. 1) and solving for the total response, we get

$$v_T = \frac{R_2}{R_1 + R_2 + KR_1} v_{s1} + \frac{R_1 R_2}{R_1 + R_2 + KR_1} i_{s2} \tag{3.8.3}$$

If we set i_{s2} to zero, we obtain the response to acting alone:

$$v_1 = \frac{R_2}{R_1 + R_2 + KR_1} v_{s1} \tag{3.8.4}$$

Similarly, if we set v_{s1} equal to zero in Equation (3. 8. 3), the response due to i_{s2} is given by:

$$v_t = \frac{R_1 R_2}{R_1 + R_2 + KR_1} i_{s2} \tag{3.8.5}$$

Comparing Equations (3. 8. 3), (3. 8. 4), and (3. 8. 5), we see that

$$v_T = v_1 + v_2$$

Thus, as expected from the superposition principle, the total response is equal to the sum of the responses for each of the independent sources acting individually.

Notice that if we zero both of the independent sources ($v_{s1} = 0$ and $i_{s2} = 0$), the response becomes zero. Hence, the dependent source does not contribute to the total response. However, the dependent source affects the contributions of the two independent sources. This is evident because the gain parameter K of the dependent source appears in the expressions for both v_1 and v_2. In general, dependent sources do not contribute a separate term to the total response, and we must not zero dependent sources in applying superposition.

3.8.1 Application of superposition principal

We can apply superposition in circuit analysis by analyzing the circuit for each source separately. Then, we add the individual responses to find the total response. Sometimes, the analysis of a circuit is simplified by considering each independent source separately. We illustrate this with an example.

Example 3.8.1 Circuit analysis using superposition

Use superposition to solve the circuit shown in Figure for example 3.8.1(a) for the voltage v_T.

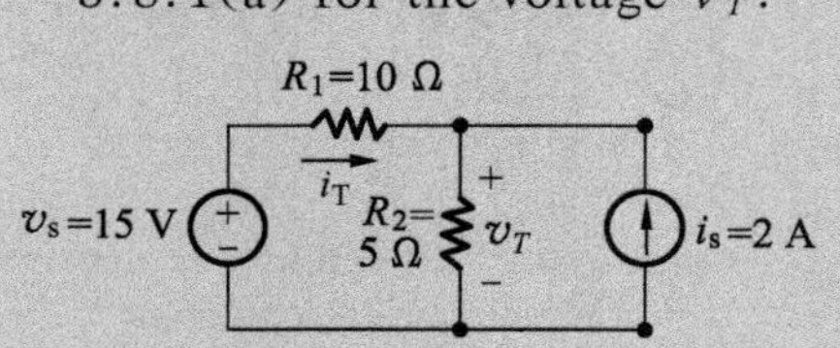

(a) Original circuit

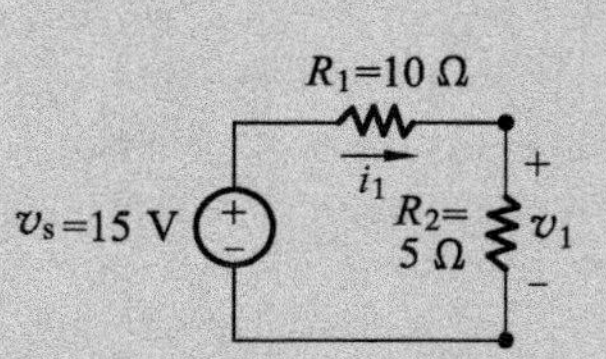

(b) Circuit with only the voltage source active

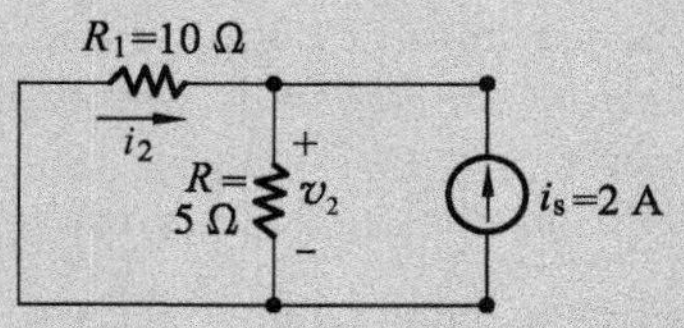

(c) Circuit with only the current source active

Figure for example 3.8.1

Solution: We analyze the circuit with only one source activated at a time and add the response. Figure for example 3.8.1(b) shows the circuit with only voltage source active, replacing the ideal current source with an open circuit deactivating it. The response can be found by applying the voltage-division principle:

$$v_1 = \frac{R_2}{R_1 + R_2} v_s = 5 \text{ V}$$

Next, we analyze the circuit with only the current source active, replacing the ideal voltage source with a short circuit deactivating it. The circuit is shown in Figure for example 3.8.1(c). In this case, the resistors R_1 and R_2 are in parallel, and the equivalent resistance is

$$R_{eq} = \frac{1}{1/R_1 + 1/R_2} = 3.33 \ \Omega$$

The voltage due to the current source is given by

$$v_2 = i_s R_{eq} = 6.66 \text{ V}$$

Finally, we obtain the total response by adding the individual responses:

$$v_T = v_1 + v_2 = 11.66 \text{ V}$$

3.8.2 Linearity

If we plot voltage versus current for a resistance, we have a straight line. This is illustrated in Figure 3.8.2.

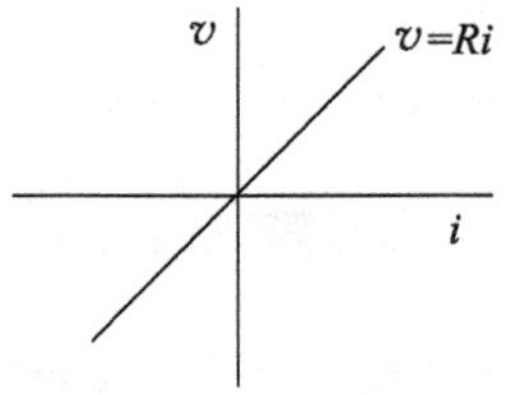

Figure 3.8.2 A resistance that obeys Ohm's law is linear

Thus, we say that Ohm's law is a **linear equation.** Similarly, the current in the controlled source shown in Figure 3.8.1 is given by $i_{cs} = ki_x$, which is also a linear equation. In this book, the term **linear controlled source** means a source whose value is a constant times a control variable that is a current or a voltage appearing in the network.

Some examples of nonlinear equations are

$$v=10i^2 \text{ or } i_{cs}=k\cos(i_x) \text{ or } i=e^v$$

The superposition principle does not apply to any circuit that has element(s) described by nonlinear equation(s). We will encounter nonlinear elements later in our study of electronic circuits.

Furthermore, superposition does not apply for power in resistances, because $P=\frac{v^2}{R}$ and $P=i^2R$ are nonlinear equations.

3.9 Simulating analysis of Thevenin's theorem using Multisim

Multisim is a software used by National Instruments Company to simulate and design electrical circuit. It has powerful function and is easy to learn.

3.9.1 Creating the circuit

Open multism10. Place components and indicators showed in Figure 3.9.1 on work space. Find J1 in the menu (place components switch, SPDT), XMM1 is in the menu [simulate instruments multimeter].

Connect components by using lines to create a circuit like Figure 3.9.1. Then connect an Ammeter and a Voltmeter to terminals a-b by J1.

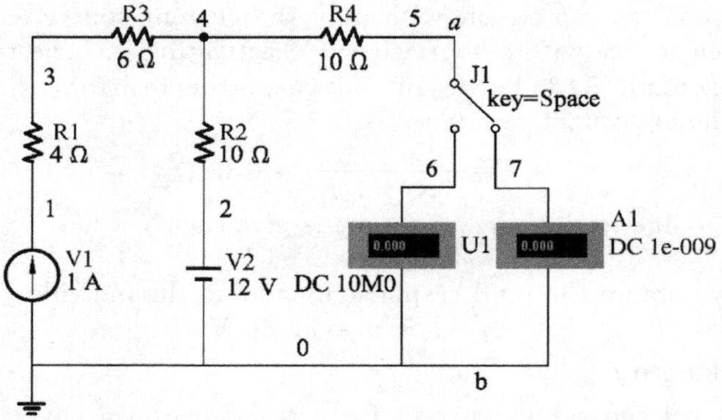

Figure 3.9.1 Creating the circuit for Thevenin theorem

3.9.2 Simulating the circuit

Press [space]. Let J1 connect to the voltmeter. Click simulate button , and get the open-circuit voltage of ports a, b. Press [space] again. Connect J1 to the Ammeter. Click simulate button , and get the short-circuit current of ports a, b.

Short the voltage source and open the current source in Figure 3.9.1, and create a passive two-terminal circuit like Figure 3.9.2. Connect a Multimeter 

between points 5 and 0. Measure the equivalent resistor.

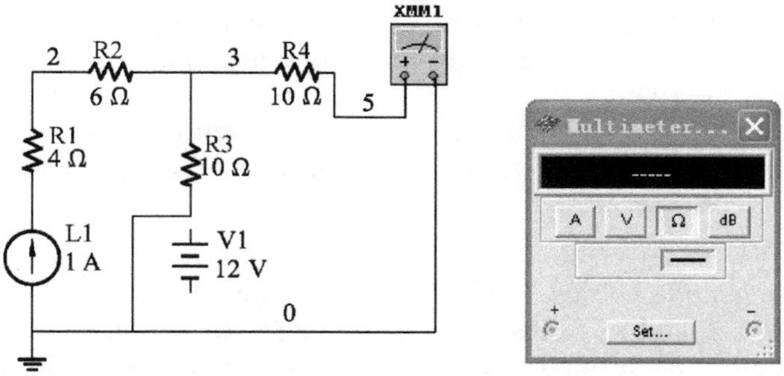

Figure 3. 9. 2 Passive two-terminal circuit

Use the calculation $R_0 = \frac{v_{oc}}{I_{sc}}$ and measurement R_0 to check if the two resistors are the same.

3.9.3 The Thevenin's equivalent circuit

Measure the current I in Figure 3. 9. 1 by connecting an Ammeter like Figure 3. 9. 3. Click simulate button , and get the data shown in the Ammeter.

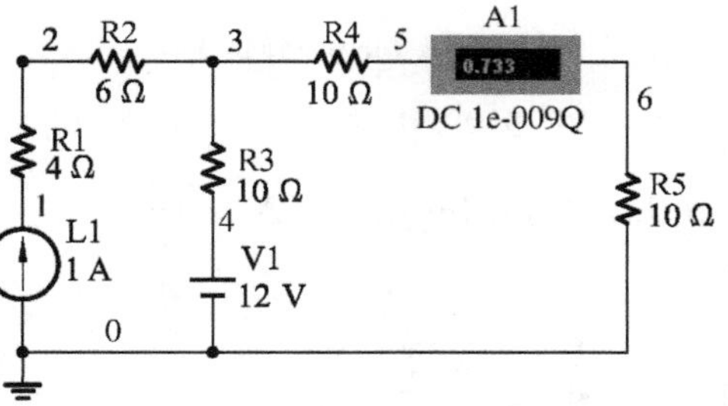

Figure 3. 9. 3 The original circuit

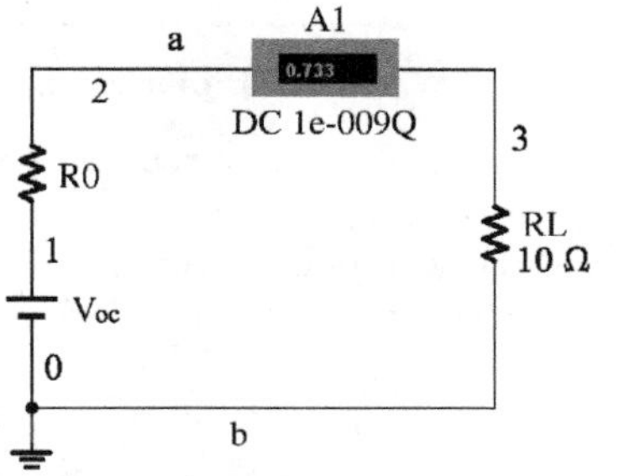

Figure 3. 9. 4 Thevenin's equivalent circuit

According to the Thevenin theorem, create the new circuit shown in Figure 3. 9. 4. Click simulate button , and get the same value of current shown in Figure 3. 9. 4. Then we can say Figure 3. 9. 4 is the equivalent circuit of Figure 3. 9. 1.

Summary

The objective of this chapter is to provide a practical introduction to the analysis of linear resistive circuits.

1. Node-voltage and mesh-current analysis. They are analogous in concept; the choice of a preferred methods depends on the specific circuit. They are generally applicable to the circuits we analyze in this book and are amenable to solution by matrix methods.

2. Thevenin and Norton equivalents. The notion of equivalent circuits is at the heart of circuit analysis. Complete mastery of the reduction of linear resistive circuits to either equivalent form is a must.

3. The principal of superposition. This is primarily a conceptual aid that may simplify the solution of circuits containing multiple sources. It is usually not an efficient method.

4. Maximum lower transfer. Equivalent circuits provide a very clear explanation of how power is transferred from a source to a load.

Problems

3.1 Using node-voltage analysis in the circuit of Figure for problem 3.1, find the currents i_1 and i_2. Let $R_1=3\ \Omega$; $R_2=1\ \Omega$; $R_3=6\ \Omega$.

3.2 Using node-voltage analysis in the circuit of Figure for problem 3.2, find the current i through the voltage source. Let $R_1=100\ \Omega$; $R_2=5\ \Omega$; $R_3=200\ \Omega$; $R_4=50\ \Omega$; $V=50$ V, $I=0.2$ A.

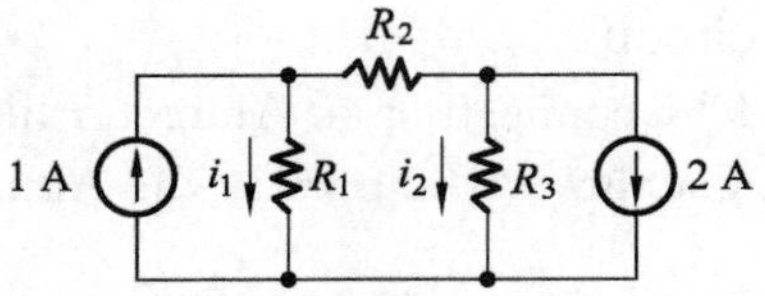

Figure for problem 3.1

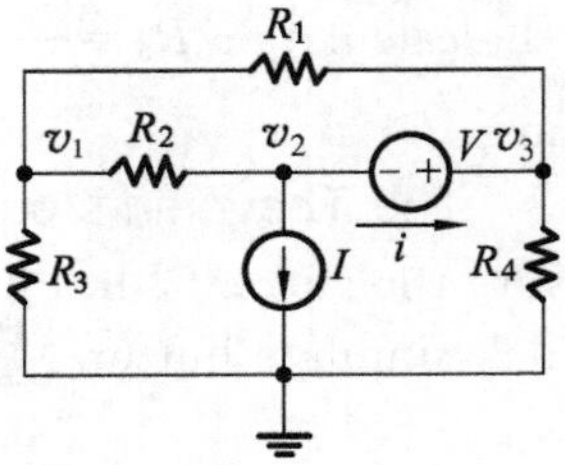

Figure for problem 3.2

3.3 (a) For the circuit shown in Figure for problem 3.3, use node-voltage method to find v_1, v_2 and i.

(b) How much power is delivered to the circuit by the 15 A source?

(c) Repeat (b) for the 5 A source.

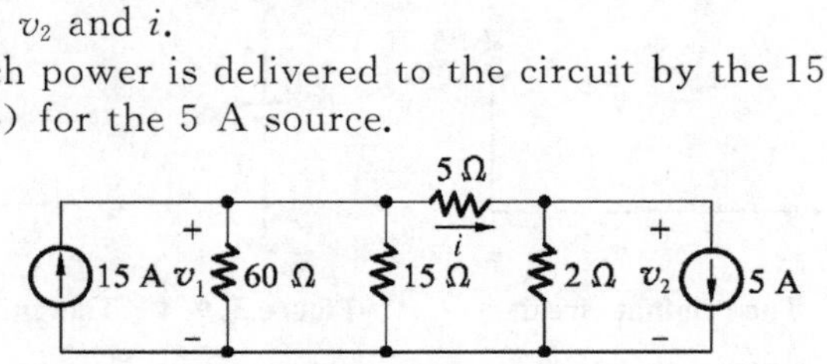

Figure for problem 3.3

3.4 Use node-voltage method to find v in the circuit shown in Figure for problem 3.4.

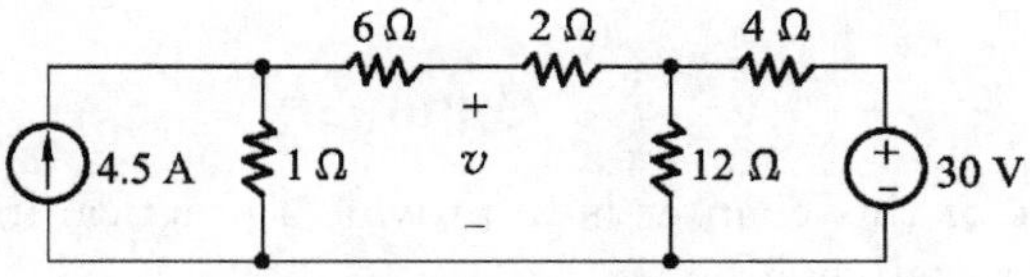

Figure for problem 3.4

3.5 (a) Use node-voltage method to find the power associated with each source in the circuit shown in Figure for problem 3.5.

(b) State whether the source is delivering power to the circuit or extracting power from the circuit.

3.6 Use node-voltage method to find v_0 in the circuit shown in Figure for problem 3.6.

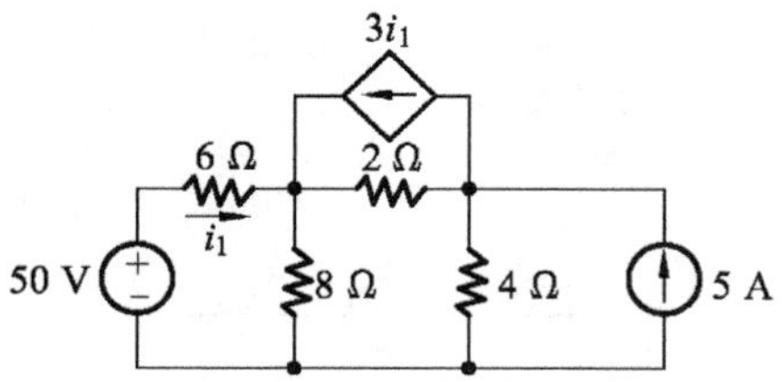

Figure for problem 3. 5

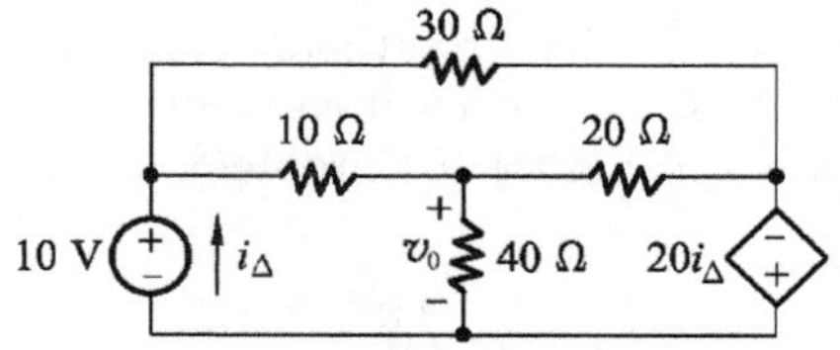

Figure for problem 3. 6

3. 7 Use node-voltage method to find i_x in the circuit shown in Figure for problem 3. 7.

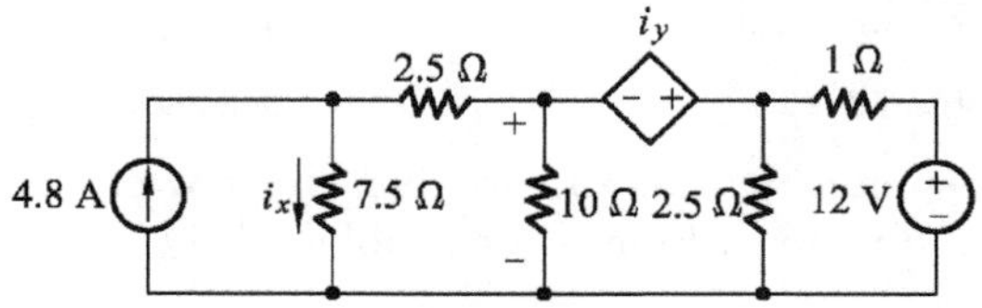

Figure for problem 3. 7

3. 8 Use node-voltage method to find v_1 in the circuit shown in Figure for problem 3. 8.

3. 9 Use mesh-current method to find

(a) the power delivered by the 80 V source in the circuit shown in Figure for problem 3. 9;

(b) the power dissipated in the 8 Ω resistor.

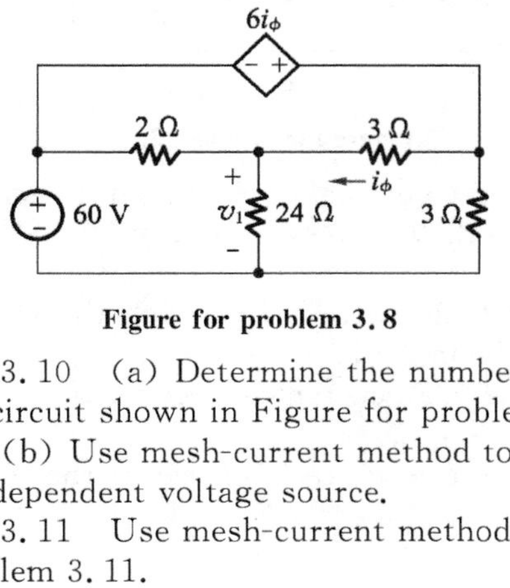

Figure for problem 3. 8

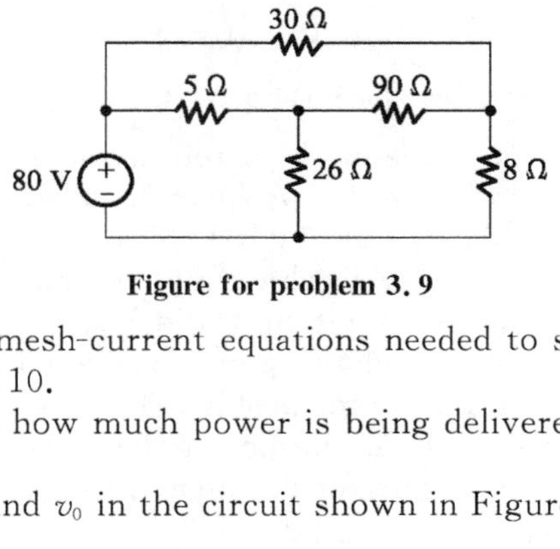

Figure for problem 3. 9

3. 10 (a) Determine the number of mesh-current equations needed to solve the circuit shown in Figure for problem 3. 10.

(b) Use mesh-current method to find how much power is being delivered to the dependent voltage source.

3. 11 Use mesh-current method to find v_0 in the circuit shown in Figure for problem 3. 11.

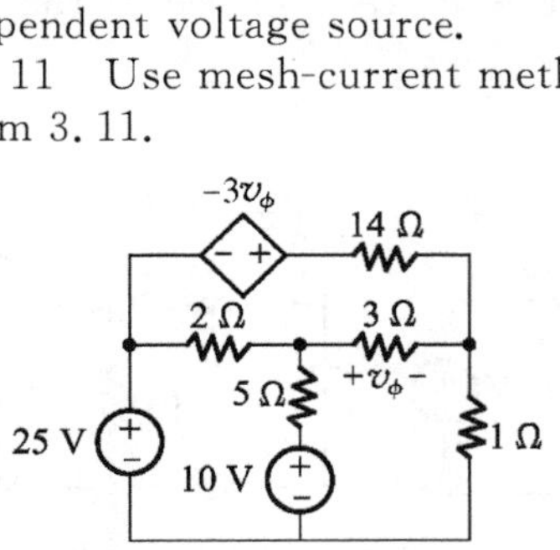

Figure for problem 3. 10

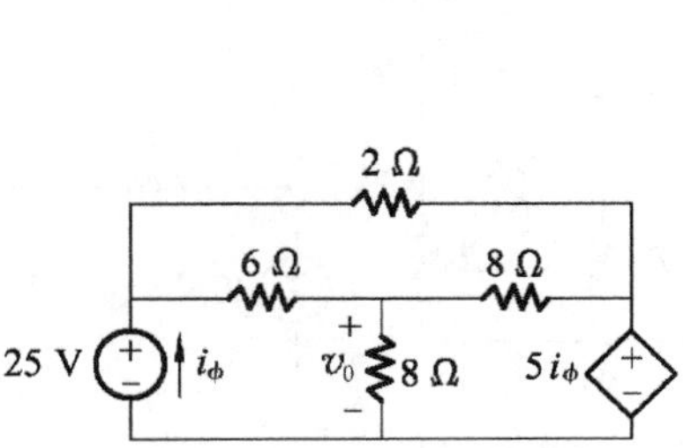

Figure for problem 3. 11

3.12 Use mesh-current method to find the power dissipated in the 2 Ω resistor in the circuit shown in Figure for problem 3.12.

3.13 Use mesh-current method to find the mesh current in the circuit i_a shown in Figure for problem 3.13.

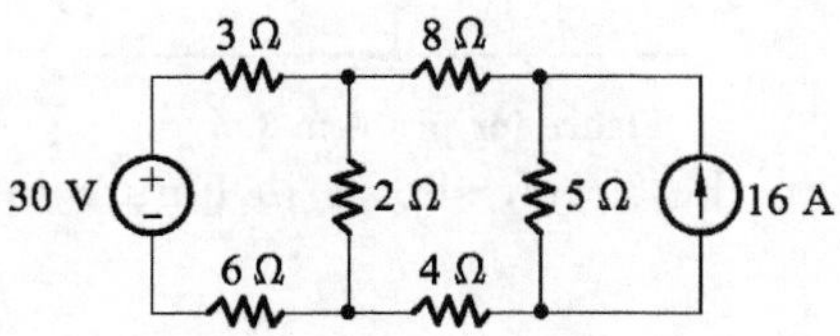

Figure for problem 3.12

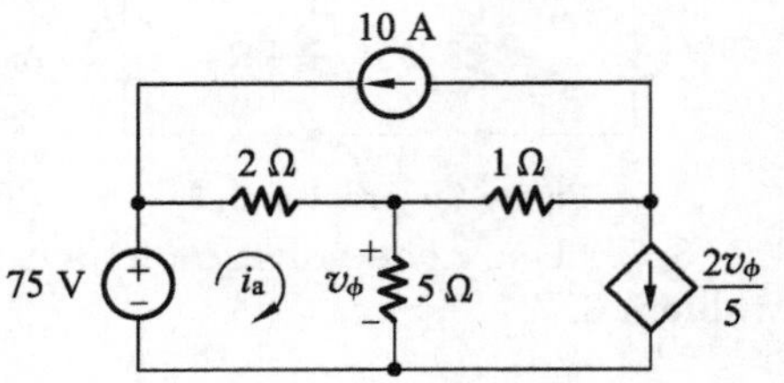

Figure for problem 3.13

3.14 Use mesh-current method to find the power dissipated in the 1 Ω resistor in the circuit shown in Figure for problem 3.14.

3.15 Use mesh-current method to find the voltage gain $A_v = \frac{v_2}{v_1}$ in the circuit shown in Figure for problem 3.15.

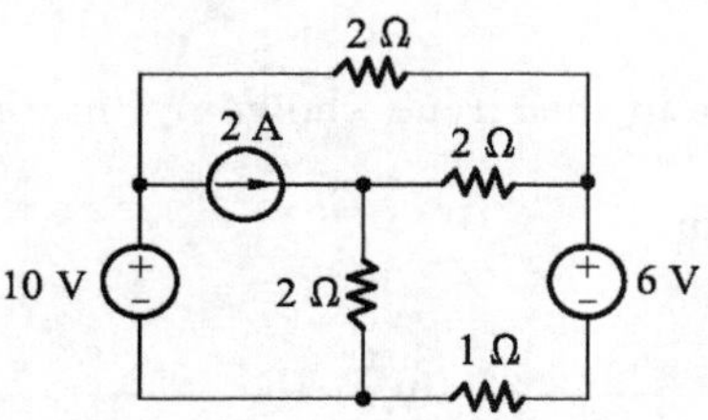

Figure for problem 3.14

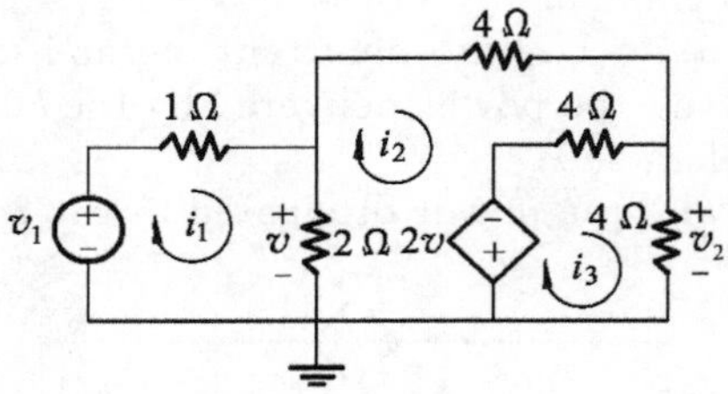

Figure for problem 3.15

3.16 Use mesh-current method to find the voltage v across R_4 in the circuit shown in Figure for problem 3.16. Let $V_{s1} = 12$ V; $V_{s2} = 5$ V; $R_1 = 50$ Ω; $R_2 = R_3 = 20$ Ω; $R_4 = 10$ Ω; $R_5 = 15$ Ω.

3.17 Use mesh-current method to find the current i through the voltage source in the circuit shown in Figure for problem 3.17.

3.18 Find the power delivered by the 2 A current source in the circuit shown in Figure for problem 3.17.

3.19 Find the power delivered by the 4 A current source in the circuit shown in Figure for problem 3.19.

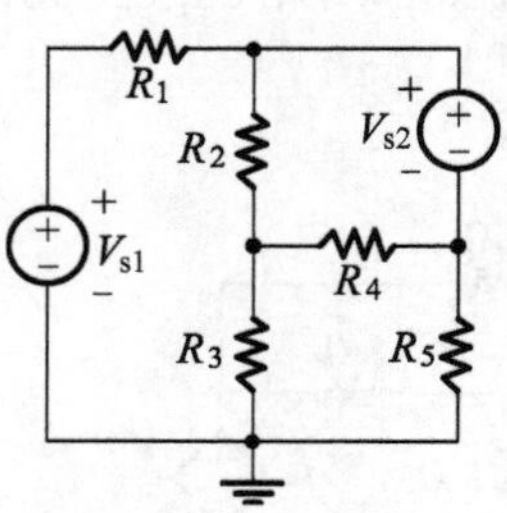

Figure for problem 3.16

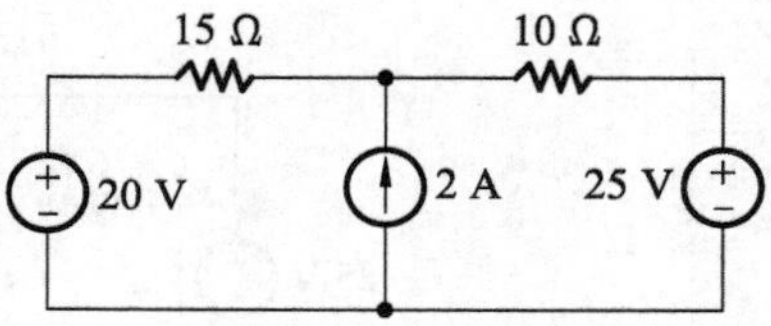

Figure for problem 3.17

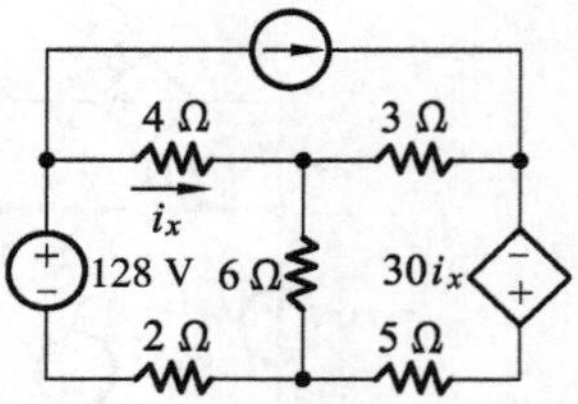

Figure for problem 3.19

3.20 Find the Thevenin equivalent circuit with respect to the terminals a, b in the circuit shown in Figure for problem 3.20.

3.21 Find the Norton equivalent circuit with respect to the terminals a, b, in the circuit shown in Figure for problem 3.21.

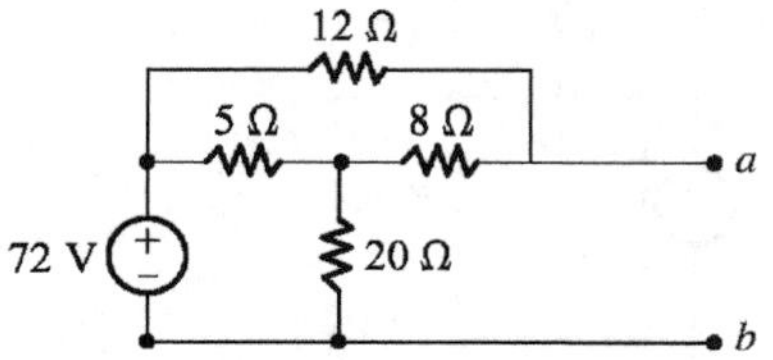

Figure for problem 3.20

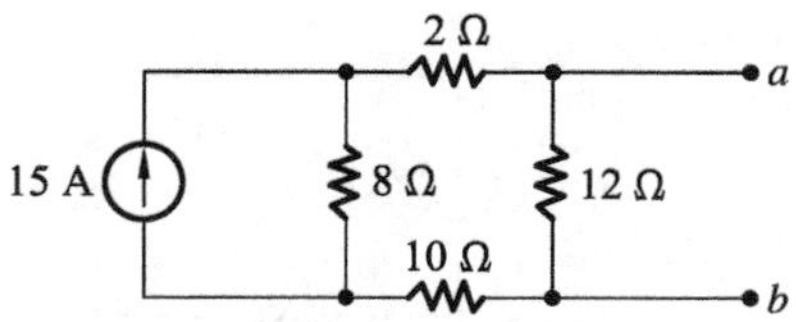

Figure for problem 3.21

3.22 A voltmeter with an internal resistance of 100 kΩ is used to measure the voltage v_{AB} in the circuit shown in Figure for problem 3.22. What is the voltmeter reading?

3.23 Find the Thevenin equivalent circuit with respect to the terminals a, b in the circuit shown in Figure for problem 3.23.

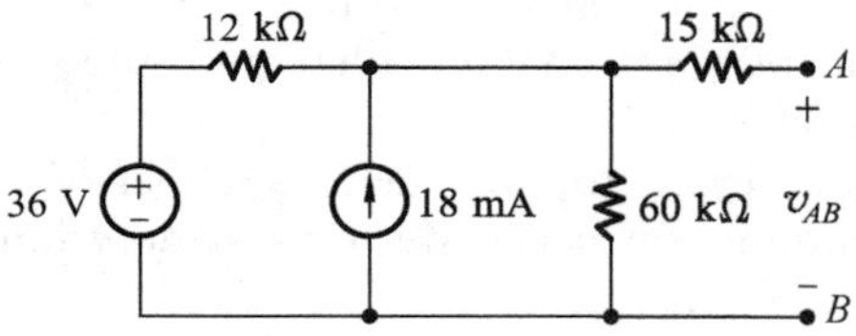

Figure for problem 3.22

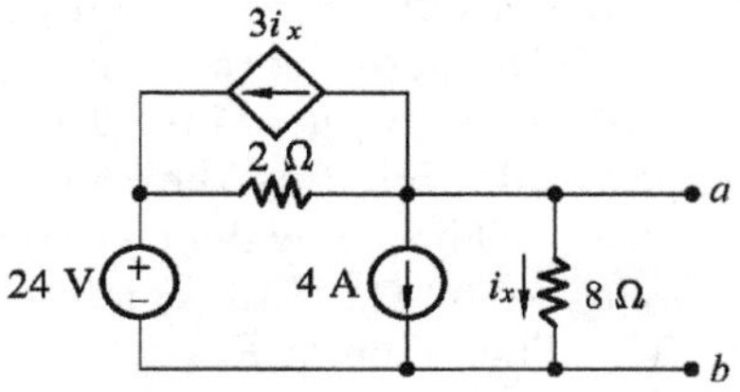

Figure for problem 3.23

3.24 Find the Thevenin equivalent circuit with respect to the terminals a, b in the circuit shown in Figure for problem 3.24.

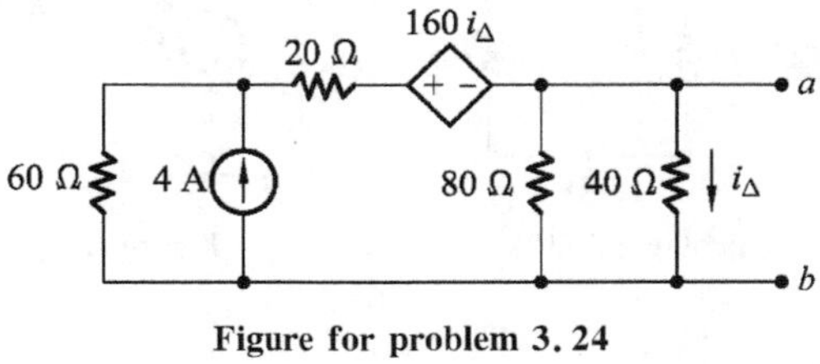

Figure for problem 3.24

3.25 Find the Norton equivalent circuit with respect to the terminals a, b in the circuit shown in Figure for problem 3.25.

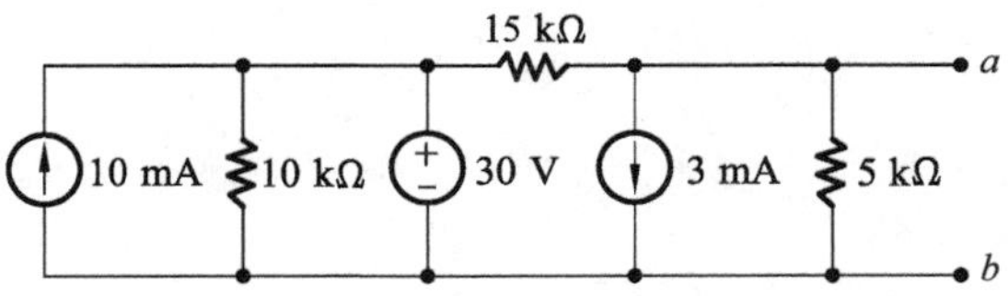

Figure for problem 3.25

3.26 Repeat problem 3.2, using the principle of superposition.

3.27 Repeat problem 3.15, using the principle of superposition.

3.28 With reference in the circuit shown in Figure for problem 3.28, determine the current through R_1, due only to the source V_{s2}.

3.29 Using superposition to find the current in the circuit shown in Figure for problem 3.29 through R_3 that is due to V_{s2}.

$$V_{s1}=V_{s2}=450\ \text{V}, R_1=7\ \Omega, R_2=5\ \Omega, R_3=10\ \Omega, R_4=R_5=1\ \Omega$$

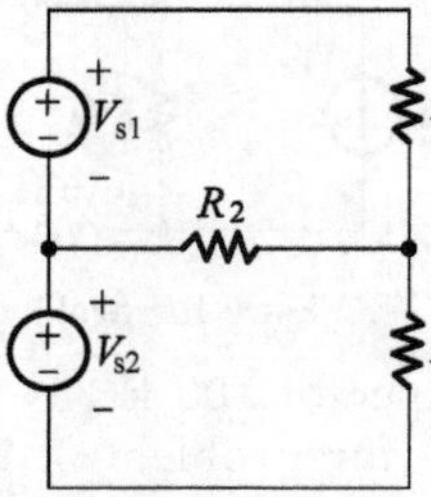

Figure for problem 3.28

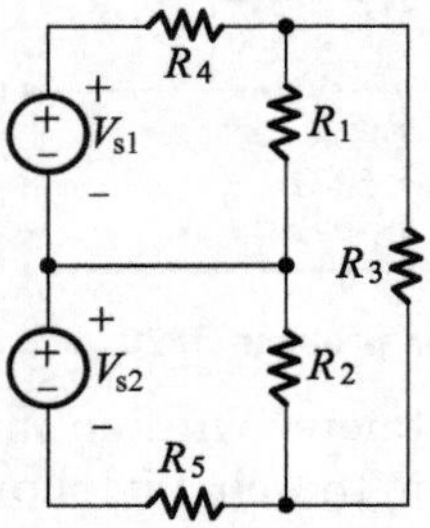

Figure for problem 3.29

3.30 (a) For the circuit shown in Figure for problem 3.30, find the value of R_L that results in maximum power being transferred to R_L.

(b) Calculate the maximum power that can be delivered to R_L.

(c) When R_L is adjusted for maximum power transfer, what percentage of the power delivered by the 360 V source reaches R_L?

3.31 The equivalent circuit shown in Figure for problem 3.31 has $V_T=$ 35 V, $R_T=600\ \Omega$. If the condition for maximum power transfer exists, determine

(a) The value of R_L.

(b) The power developed in R_L.

(c) The efficiency of the circuit.

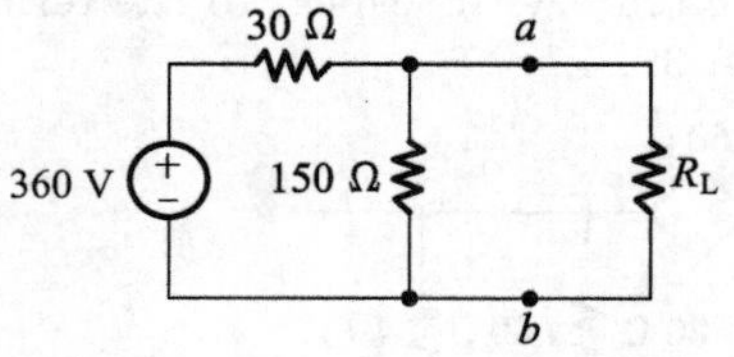

Figure for problem 3.30

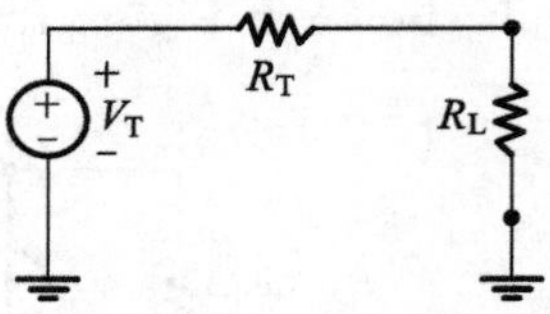

Figure for problem 3.31

CHAPTER 4

RESPONSE OF FIRST-ORDER RL AND RC CIRCUITS

Introduction

This chapter is to develop a systematic methodology for the solution of first-order circuits excited by switch DC sources. The chapter presents a unified approach to determining the transient response of linear RC, RL, and RLC circuits; and although the methods presented in the chapter focus only on first-order circuits, the approach to the transient solution is quite general.

4.1 The inductor

An inductor is usually constructed by coiling a wire around some type of form. Several examples of practical construction are illustrated in Figure 4. 1. 1. Current flowing through the coil creates a magnetic field or flux that links the coil. Frequently, the coil form is composed of a magnetic material such as iron or iron oxides that increases the magnetic flux for a given current. (Iron cores are often composed of thin sheets called laminations.)

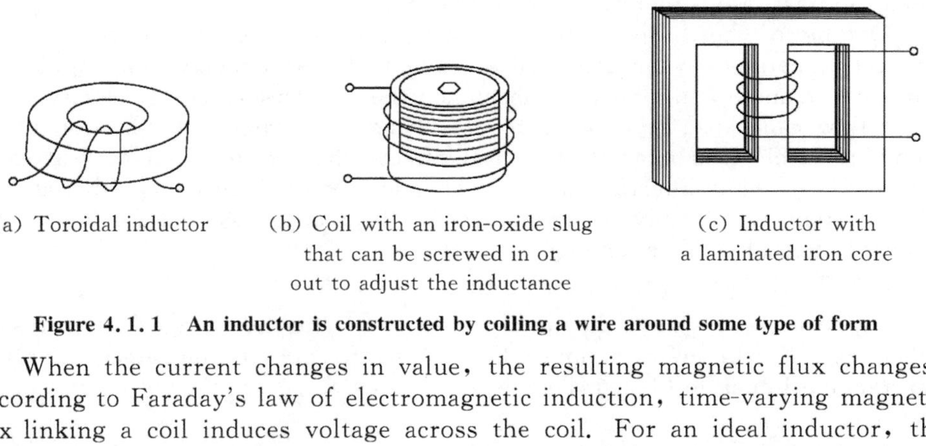

(a) Toroidal inductor (b) Coil with an iron-oxide slug that can be screwed in or out to adjust the inductance (c) Inductor with a laminated iron core

Figure 4. 1. 1 An inductor is constructed by coiling a wire around some type of form

When the current changes in value, the resulting magnetic flux changes. According to Faraday's law of electromagnetic induction, time-varying magnetic flux linking a coil induces voltage across the coil. For an ideal inductor, the voltage is proportional to the time rate of change of the current. Furthermore, the polarity of the voltage is such as to oppose the change in current. The constant of proportionality is called inductance, usually denoted by the letter L.

4.1.1 Ideal inductor

The circuit symbol for inductance is shown in Figure 4. 1. 2. In equation form, the voltage and current are related by Equation(4. 1. 1).

$$v(t)=L\frac{di}{dt} \tag{4.1.1}$$

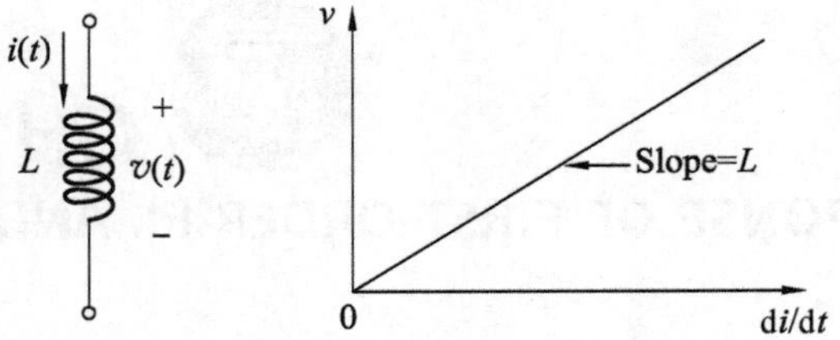

Figure 4.1.2 Circuit symbol and the v-i relationship for inductance

As usual, we have assumed the passive reference configuration. In case the references are opposite to the passive configuration, Equation (4.1.1) becomes

$$v(t)=-L\frac{di}{dt} \tag{4.1.2}$$

Inductance has unit of henries (H), which are equivalent to volt seconds per ampere. Typically, we deal with inductances ranging from a fraction of a microhenry (μH) to several tens of henries.

4.1.2 Fluid-flow Analogy

The fluid-flow analogy for inductance is the inertia of the fluid flowing through a frictionless pipe of constant diameter. The pressure differential between the ends of the pipe is analogous to voltage, and the flow rate or velocity is analogous to current. Thus, the acceleration of the fluid is analogous to rate of change of current. A pressure differential exists between the ends of the pipe only when the flow rate is increasing or decreasing.

One place where the inertia of flowing fluid is encountered when a valve (typically operated by an electrical solenoid) closes suddenly, cutting off the flow. For example, in a washing machine, the sudden change in velocity of the water flow can cause high pressure, resulting in a bang and vibration of the plumbing. This is similar to electrical effects that occur when current in an inductor is suddenly interrupted. An application for the high voltage that appears when current is suddenly interrupted is in the ignition system for a gasoline-powered internal combustion engine.

4.1.3 Current in terms of voltage

Suppose that we know the initial current $i(t_0)$ and the voltage $v(t_0)$ across an inductance. Furthermore, suppose that we need to compute the current for $t>t_0$. Rearranging Equation (4.1.1), we have

$$di=\frac{1}{L}v(t)dt \tag{4.1.3}$$

Integrating both sides, we find that

$$\int_{i(t_0)}^{i(t)} di = \frac{1}{L}\int_{t_0}^{t} v(t)dt \tag{4.1.4}$$

Notice that the integral on the right-hand side of Equation (4.1.4) is with respect to time. Furthermore, the limits are the initial time t_0 and the time variable t. The integral on the left-hand side is with respect to current with limits that correspond to the time limits on the right-hand side. Integrating, evaluating, and rearranging, we have

$$i(t) = \frac{1}{L}\int_{t_0}^{t} v(t)\,\mathrm{d}t + i(t_0) \tag{4.1.5}$$

Notice that as long as $v(t)$ is finite, $i(t)$ can change only by an incremental amount in a time increment. Thus, $i(t)$ must be continuous with no instantaneous jumps in value.

4.1.4 Stored energy

Assuming that the references have the passive configuration, we compute the power delivered to a circuit element by taking the product of the current and the voltage:

$$p(t) = v(t)i(t) \tag{4.1.6}$$

Using Equation (4.1.1) to substitute for the voltage, we obtain

$$p(t) = Li(t)\frac{\mathrm{d}i}{\mathrm{d}t} \tag{4.1.7}$$

Consider an inductor having an initial current $i(t) = 0$. Then, the initial electrical energy stored is zero. Furthermore, assume that between time t_0 and some later time t, the current changes from 0 to $i(t)$. As the current magnitude increases, energy is delivered to the inductor, where it is stored in the magnetic field.

Integrating the power from t_0 to t, we find the energy delivered:

$$w(t) = \int_{t_0}^{t} p(t)\,\mathrm{d}t \tag{4.1.8}$$

Using Equation 4.1.7 to substitute for power, we have

$$w(t) = \int_{t_0}^{t} Li(t)\frac{\mathrm{d}i}{\mathrm{d}t}\mathrm{d}t \tag{4.1.9}$$

Canceling differential time and changing the limits to the corresponding current, we get

$$w(t) = \int_{0}^{i(t)} Li\,\mathrm{d}i \tag{4.1.10}$$

Integrating and evaluating, we obtain

$$w(t) = \frac{1}{2}Li^2(t) \tag{4.1.11}$$

This represents energy stored in the inductance that is returned to the circuit if the current changes back to zero.

We should note the following important properties of an inductor.

(1) Note from Equation (4.1.2) that the voltage across an inductor is zero when the current is constant. Thus,

an inductor acts like a short circuit to DC.

(2) An important property of the inductor is its opposition to the change in current flowing through it.

The current through an inductor cannot change instantaneously.

According to Equation (4.1.2), a discontinuous change in the current through an inductor requires an infinite voltage, which is not physically possible. Thus, an inductor opposes an abrupt change in the current through it. For example, the current through an inductor may take the form shown in Figure 4.1.3(a), whereas the inductor current cannot take the form shown in Figure

4.1.3(b) in real-life situations due to the discontinuities.

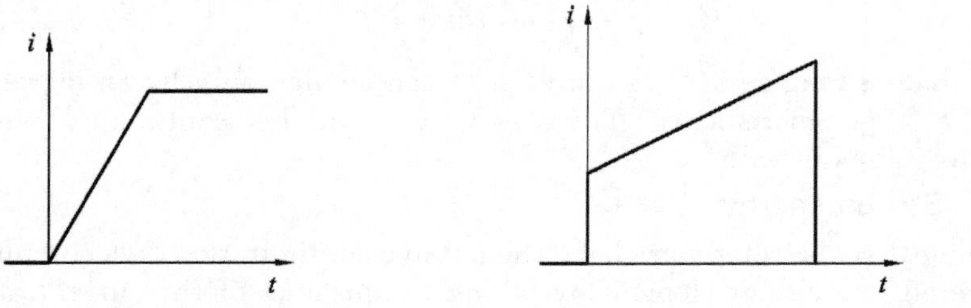

(a) Allowed (b) Not allowed; an abrupt change is not possible

Figure 4.1.3 Current through an inductor

(3) The ideal inductor does not dissipate energy. The energy stored in it can be retrieved at a later time. The inductor takes power from the circuit when storing energy and delivers power to the circuit when returning previously stored energy.

Example 4.1.1 Voltage, power, and energy for an inductance

The current through a 5 H inductance is shown in Figure for example 4.1.1(a). Plot the voltage, power and stored energy to scale versus time for t between 0 and 5 s.

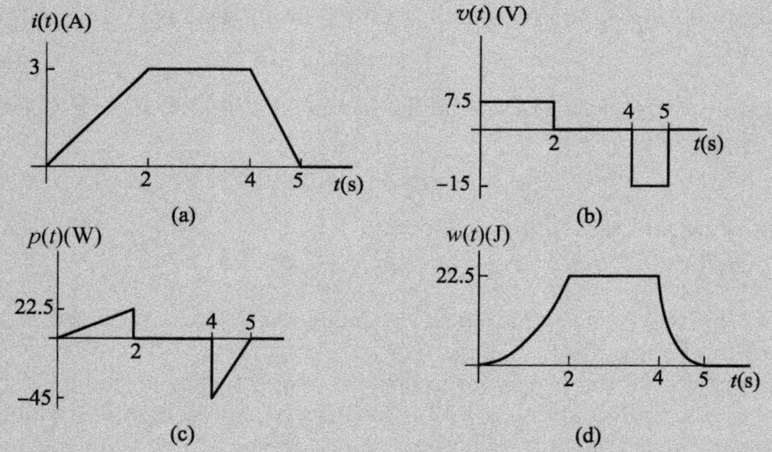

Figure for example 4.1.1

Solution: We use Equation (4.1.1) to compute voltages:

$\frac{di}{dt}=1.5$ A/s when $0<t<2$ s, then $v=7.5$ V

$\frac{di}{dt}=0$ when $2<t<4$ s, then $v=0$

$\frac{di}{dt}=-3$ A/s when $4<t<5$ s, then $v=-15$ V

A plot of the voltage versus time is shown in Figure for example 4.1.1(b).

Next, we obtain power by taking the product of current and voltage at each point of time. The resulting plot is shown in Figure for example 4.1.1(c).

Finally, we use Equation (4.1.11) to compute the stored energy as a function of time. The resulting plot is shown in Figure for example 4.1.1(d).

Through the example 4.4.4 we notice that as current magnitude increases, power is positive and stored energy accumulates. When the current is constant, the voltage is zero, the power is zero, and the stored energy is constant. When the current magnitude falls toward zero, the power is negative, showing that energy is being returned to the other parts of the circuit.

4.1.5 Inductors in series and in parallel

It can be shown that the equivalent inductance for a series circuit is equal to the sum of the inductances connected in series. On the other hand, for inductances in parallel, we find the equivalent inductance by taking the reciprocal of the sum of the reciprocals of the parallel inductances. Series and parallel equivalents for inductances are illustrated in Figure 4.1.4. Notice that inductances are combined in exactly the same way as resistors. These facts can be proven by following the pattern used earlier in this chapter to derive the equivalents for series capacitances.

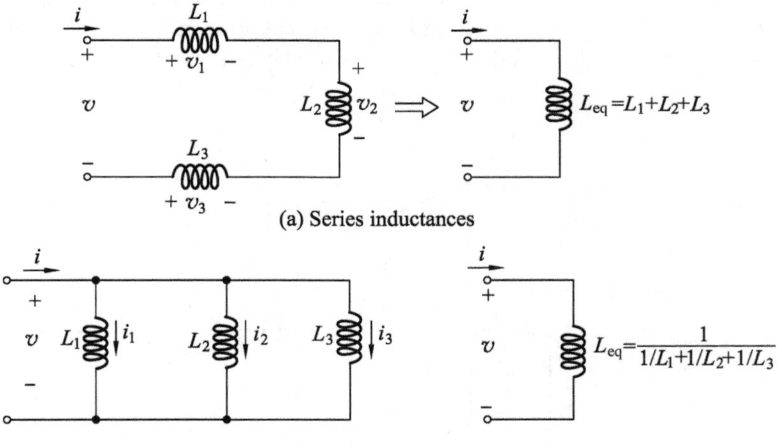

(b) Parallel inductances

Figure 4.1.4 Inductors in series and parallel are combined in the same manner as resistors

4.2 The capacitor

A capacitor is a passive element designed to store energy in its electric field. Besides resistors, capacitors are the most common electrical components. Capacitors are used extensively in electronics, communications, computers, and power systems. For example, they are used in the tuning circuits of radio receivers and as dynamic memory elements in computer systems.

4.2.1 Ideal capacitor

A capacitor is typically constructed as depicted in Figure 4.2.1. A capacitor consists of two conducting plates separated by an insulator (or dielectric).

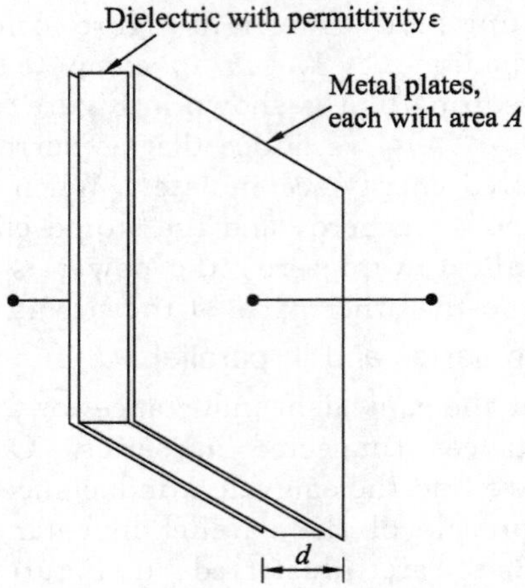

Figure 4.2.1 A typical capacitor

When a voltage source v is connected to the capacitor, as shown in Figure 4.2.2, the source deposits a positive charge q on one plate and a negative charge $-q$ on the other. The capacitor is said to store the electric charges. The amount of charges stored, represented by q, is directly proportional to the applied voltage, so that

$$q = Cv \tag{4.2.1}$$

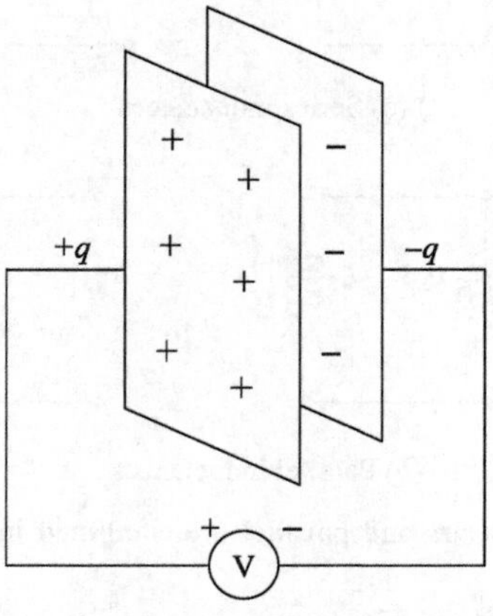

Figure 4.2.2 A capacitor with applied voltage

Where C, the constant of proportionality, is known as the capacitance of the capacitor. The unit of capacitance is the farad (F).

Capacitors are commercially available in different values and types. Typically, capacitors have values in the picofarad (pF) to microfarad (μF) range. They are described by the dielectric material they are made of and by whether they are fixed or a variable type. Figure 4. 2. 3 shows the circuit symbols for fixed and variable capacitors. Note that according to the passive sign convention, current is considered to flow into the positive terminal of the capacitor when the capacitor is being charged, and out of the positive terminal when the capacitor is discharging.

(a) fixed capacitor (b) variable capacitor

Figure 4. 2. 3 Circuit symbols

Figure 4. 2. 4 shows common types of fixed-value capacitors. Polyester capacitors are light in weight, stable, and their change with temperature is predictable. Instead of polyester, other dielectric materials such as mica and polystyrene may be used. Film capacitors are rolled and housed in metal or plastic films. Electrolytic capacitors produce very high capacitance. Figure 4. 2. 5 shows the most common types of variable capacitors.

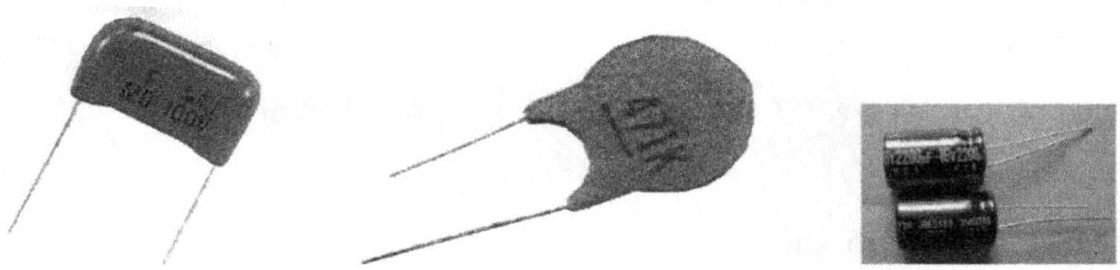

(a) Polyester capacitor (b) Ceramic capacitor (c) Electrolytic capacitor

Figure 4. 2. 4 Fixed capacitors

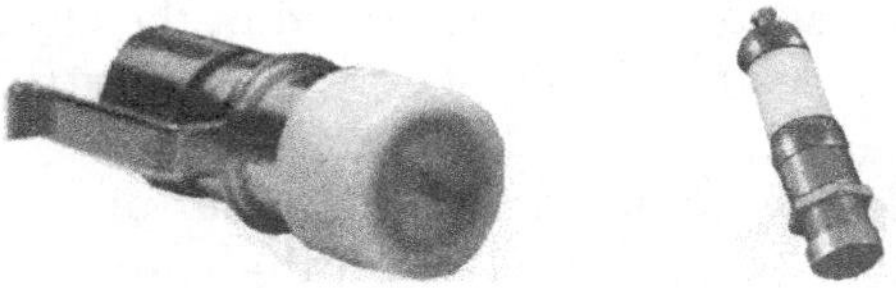

(a) Trimmer Capacitor (b) Filmtrim Capacitor

Figure 4. 2. 5 Variable capacitor

Since

$$i=\frac{dq}{dt}$$

The current-voltage relationship of the capacitor can be obtained:

$$i=C\frac{dv}{dt} \tag{4. 2. 2}$$

This is the current-voltage relationship for a capacitor, assuming the passive sign convention. The relationship is illustrated in Figure 4. 2. 6 for a capacitor whose capacitance is independent of voltage. Capacitors that satisfy Equation (4. 2. 2) are said to be liner. For a nonlinear capacitor, the plot of the current-voltage relationship is not a straight line. Although some capacitors are nonlinear, most are liner. We will learn liner capacitors in this book.

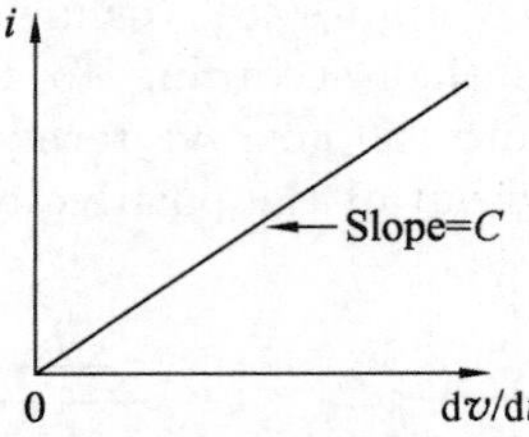

Figure 4.2.6　Current-voltage relationship of a capacitor

The voltage-current relation of the capacitor can be obtained by integrating both sides of Equation (4.2.2). We get

$$v = \frac{1}{C}\int_{-\infty}^{t} i\mathrm{d}t \tag{4.2.3}$$

or

$$v = \frac{1}{C}\int_{0}^{t} i\mathrm{d}t + v(t_0) \tag{4.2.4}$$

where $v(t_0)$ is the voltage across the capacitor at time t_0. Equation (4.2.4) shows that capacitor voltage depends on the past history of the capacitor current. Hence, the capacitor has memory property that is often exploited.

4.2.2 Stored energy

The instantaneous power delivered to the capacitor is

$$p = vi = Cv\frac{\mathrm{d}v}{\mathrm{d}t} \tag{4.2.5}$$

The energy stored in the capacitor is

$$w = \int_{-\infty}^{t} p\mathrm{d}t = C\int_{-\infty}^{t} v\frac{\mathrm{d}v}{\mathrm{d}t}\mathrm{d}t = C\int_{-\infty}^{t} v\mathrm{d}v = \frac{1}{2}Cv^2 \tag{4.2.6}$$

We note that $v(-\infty)=0$, because the capacitor is uncharged at $t=-\infty$.

Equation (4.2.6) represents the energy stored in the electric field that exists between the plates of the capacitor. This energy can be retrieved, since an ideal capacitor cannot dissipate energy.

We should note the following important properties of a capacitor:

(1) Note from Equation (4.2.2) that when the voltage across a capacitor is not changing with time (i.e. DC voltage), the current through the capacitor is zero. Thus,

a capacitor is an open circuit to DC.

(2) The voltage on the capacitor must be continuous. Thus,

the voltage on the capacitor cannot change abruptly.

The capacitor resists an abrupt change in the voltage across it. According to Equation (4.2.2), a discontinuous change in voltage requires an infinite current, which is physically impossible. For example in Figure 4.2.7(a), whereas it is not physically possible for the capacitor voltage to take the form shown in Figure 4.2.7(b), because of the abrupt changes. Conversely, the current through a capacitor can change instantaneously.

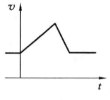

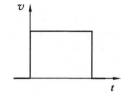

(a) Allowed (b) Not allowable; an abrupt change is not possible

Figure 4.2.7 Voltage across a Capacitor

(3) The ideal capacitor does not dissipate energy. It takes power from the circuit when storing energy in its field and returns previously stored energy when delivering power to the circuit.

4.2.3 Capacitors in series and in parallel

In order to obtain the equivalent capacitor of n capacitors in parallel, consider the circuit in Figure 4.2.8(a). The equivalent circuit is in Figure 4.2.8(b). Note that the capacitor have the same voltage v across them. Apply KCL to Figure 4.2.8(a)

$$i=i_1+i_2+i_3+\cdots+i_n \tag{4.2.7}$$

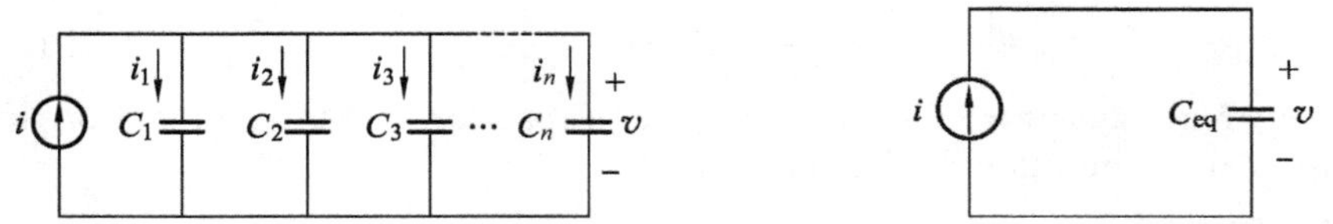

(a) Parallel-connected n capacitors (b) Equivalent circuit for the parallel capacitors

Figure 4.2.8

But $i_k=C_k\dfrac{\mathrm{d}v}{\mathrm{d}t}$, hence,

$$\begin{aligned} i &= C_1\frac{\mathrm{d}v}{\mathrm{d}t}+C_2\frac{\mathrm{d}v}{\mathrm{d}t}+C_3\frac{\mathrm{d}v}{\mathrm{d}t}+\cdots C_n\frac{\mathrm{d}v}{\mathrm{d}t} \\ &= \left(\sum_{k=1}^{n}C_k\right)\frac{\mathrm{d}v}{\mathrm{d}t}=C_{\mathrm{eq}}\frac{\mathrm{d}v}{\mathrm{d}t} \end{aligned} \tag{4.2.8}$$

where

$$C_{\mathrm{eq}}=C_1+C_2+C_3+\cdots+C_n \tag{4.2.9}$$

The equivalent capacitor of n parallel connected capacitor is the sum of the individual capacitances.

We observe that capacitors in parallel combine in the same manner as resistors in series.

We now obtain the equivalent capacitor of n capacitors in series by comparing the circuit in Figure 4.2.9 (a) with the equivalent circuit in Figure 4.2.9(b). Note that the current flows through the capacitors. Apply KVL to the loop in Figure 4.2.9(a)

$$v=v_1+v_2+v_3+\cdots+v_n \tag{4.2.10}$$

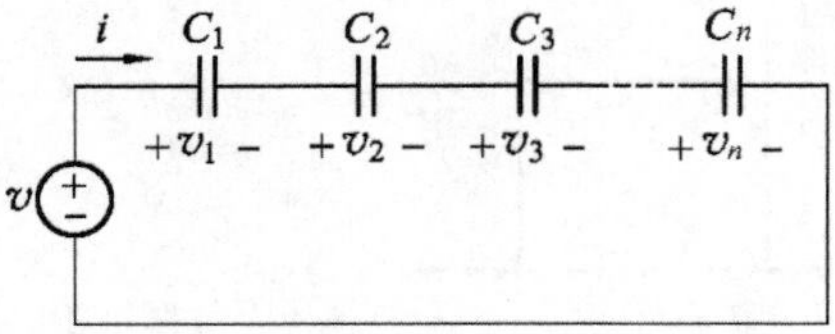

(a) Serious-connected n capacitors

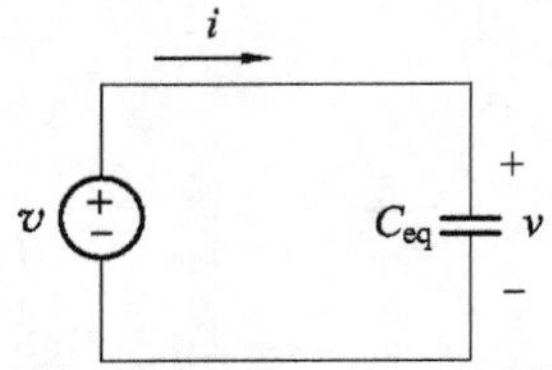

(b) Equivalent circuit for the serious capacitors

Figure 4.2.9

But $v=\frac{1}{C}\int_{t_0}^{t} i(t)$, therefore,

$$\begin{aligned} v &= \frac{1}{C_1}\int_{t_0}^{t} i(t) + \frac{1}{C_2}\int_{t_0}^{t} i(t) + \frac{1}{C_3}\int_{t_0}^{t} i(t) + \cdots + \frac{1}{C_n}\int_{t_0}^{t} i(t) \\ &= \left(\frac{1}{C_1}+\frac{1}{C_2}+\frac{1}{C_3}+\cdots+\frac{1}{C_n}\right)\int_{t_0}^{t} i(t) \\ &= \frac{1}{C_{eq}}\int_{t_0}^{t} i(t) \end{aligned} \tag{4.2.11}$$

where

$$\frac{1}{C_{eq}}=\frac{1}{C_1}+\frac{1}{C_2}+\frac{1}{C_3}+\cdots+\frac{1}{C_n} \tag{4.2.12}$$

The equivalent capacitor of series-connected capacitor is the reciprocal of the sum of reciprocal of the individual capacitances.

Example 4.2.1

(a) Calculate the charge stored in a 3 pF capacitor with 20 V across it.
(b) Find the energy stored in the capacitor.

Solution: (a) Since

$$q=Cv$$

$$q=3\times10^{-12}\times20=60\ \text{pC}$$

(b) The energy stored is

$$w=\frac{1}{2}Cv^2=\frac{1}{2}\times3\times10^{-12}\times400=600\ \text{pJ}$$

Example 4.2.2

The voltage across a 5 μF capacitor is $v(t)=10\cos 6\,000t$ V; Calculate the current through it.

Solution: By definition, the current is

$$\begin{aligned} i(t) &= C\frac{\mathrm{d}v}{\mathrm{d}t}=5\times10^{-6}\frac{\mathrm{d}}{\mathrm{d}t}(10\cos 6\,000t) \\ &= -5\times10^{-6}\times6\,000\times10\sin 6\,000t=-0.3\sin 6\,000t\ \text{A} \end{aligned}$$

4.3 The natural response of an RC circuit

A source-free circuit occurs when its DC source is suddenly disconnected. The energy already stored in the capacitor is released to the resistors.

4.3.1 Deriving the expression for the voltage

Consider a series combination of a resistor and initially charged capacitor, as shown in Figure 4.3.1. We assume the voltage $v(t)$ across the capacitor. Since the capacitor is initially charged, we can assume that at time $t=0$, the initial voltage is

$$v(0)=V_0 \tag{4.3.1}$$

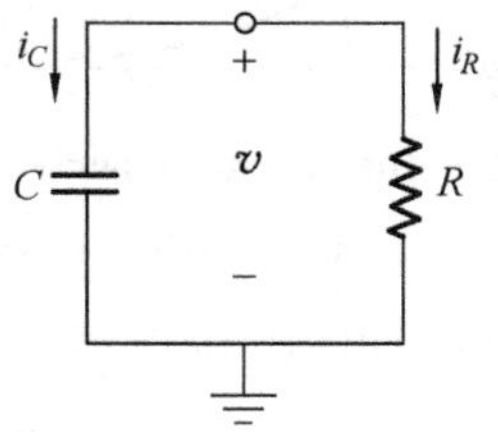

Figure 4.3.1 A source-free RC circuit

Apply KCL at the top node of the circuit in Figure 4.3.1

$$i_C+i_R=0 \tag{4.3.2}$$

By definition $i_C=C\frac{dv}{dt}$, and $i_R=\frac{v}{R}$. Thus,

$$C\frac{dv}{dt}+\frac{v}{R}=0 \tag{4.3.3}$$

This is a first-order differential equation, since only the first derivative of v is involved.

$$\frac{dv}{v}=-\frac{1}{RC}dt \tag{4.3.4}$$

Integrating both sides, we get

$$\ln v=-\frac{t}{RC}+\ln A$$

where $\ln A$ is the integration constant. Thus,

$$\ln\frac{v}{A}=-\frac{t}{RC} \tag{4.3.5}$$

Taking powers of e produces

$$v(t)=Ae^{\frac{-t}{RC}} \tag{4.3.6}$$

But from the initial conditions, $v(0)=A=V_0$. Hence,

$$v(t)=V_0e^{\frac{-t}{RC}} \tag{4.3.7}$$

This shows that the voltage response of the RC circuit is an exponential decay of the initial voltage. Since the response is due to the initial energy stored and the physical characteristics of the circuit and not due to some external voltage or current source, it is called the natural response of the circuit.

The natural response of a circuit refers to the behavior (in terms of voltage and currents) of the circuit itself, with no external sources of excitation.

4.3.2 The significance of the time constant

The natural response is illustrated graphically in Figure 4.3.2. Note that at $t=0$, we have the correct initial condition as in Equation (4.3.1). As t increases, the voltage decreases toward zero. The rapidity with which the voltage decreases is expressed in terms of the **time constant**, denoted by τ.

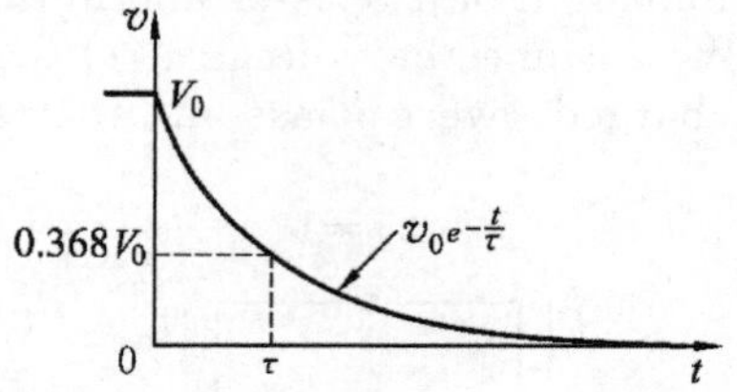

Figure 4.3.2 The voltage response of RC circuit

The time constant of a circuit is the time required for the response to decay by a factor of $\frac{1}{e}$ or 36.8 percent of its initial value. So Equation (4.3.7) becomes

$$V_0 e^{\frac{-\tau}{RC}}=V_0 e^{-1}=0.368V_0$$

or

$$\tau=RC \qquad (4.3.8)$$

In terms of the time constant Equation (4.3.7) can be written as

$$v(t)=V_0 e^{\frac{-t}{\tau}} \qquad (4.3.9)$$

With a calculator it is easy to show that the value of $\frac{v(t)}{V_0}$ is as shown in Table 4.3.1. It is evident from Table 4.3.1 that the voltage $v(t)$ is less than 1 percent of V_0 after 5τ (five time constants). Thus, it is customary to assume that the capacitor is fully discharged (or charged) after five time constants. In other words, it takes 5τ for the circuit to reach its final state or steady state when no charge takes place with time. Notice that for every time interval of t, the voltage is reduced by 36.8% of its previous value, regardless of the value of τ.

Table 4.3.1 Values of $\frac{v(t)}{V_0}=e^{\frac{-t}{\tau}}$

t	τ	2τ	3τ	4τ	5τ
$e^{-\frac{t}{\tau}}$	e^{-1}	e^{-2}	e^{-3}	e^{-4}	e^{-5}
$\frac{v(t)}{V_0}$	0.367 88	0.135 34	0.049 79	0.018 32	0.006 74

Observe from Equation (4.3.9) that the smaller the time constant is, the more rapidly the voltage decreases, that is, the faster the response is. This is illustrated in Figure 4.3.3. A circuit with a small time constant gives a fast response in that it reaches the steady state (or final state) quickly due to quick dissipation of energy stored, whereas a circuit with a large time constant gives a slow response because it takes longer to reach steady state in five time constants.

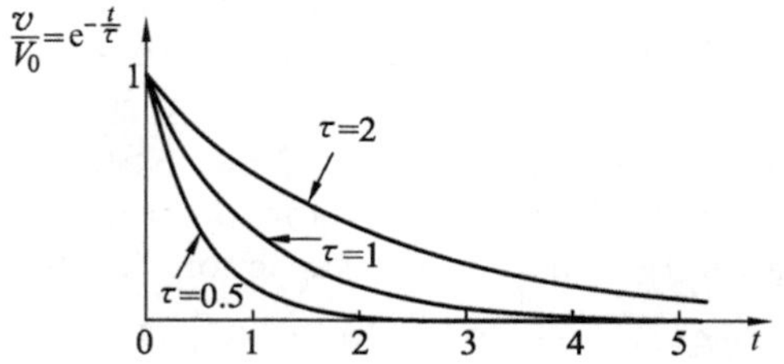

Figure 4.3.3 Plot of for $\frac{v(t)}{V_0}=e^{-\frac{t}{\tau}}$ various values of the time constant

With the voltage $v(t)$ in Equation (4.3.9), we can find the current $i_R(t)$,

$$i_R(t)=\frac{v(t)}{R}=\frac{V_0}{R}e^{-\frac{t}{\tau}} \tag{4.3.10}$$

The power dissipated in the resistor is

$$p(t)=vi_R=\frac{V_0^2}{R}e^{\frac{-2t}{\tau}} \tag{4.3.11}$$

The key to working with a source-free RC circuit is finding:

(1) the initial voltage $v(0)=V_0$ across the capacitor;

(2) the time constant τ.

With these two items, we obtain the response as the capacitor voltage $v_C(t)=v(t)=v(0)e^{-\frac{t}{\tau}}$. Once the capacitor voltage is first obtained, other variables (capacitor current i_C, resistor voltage v_R, and resistor current i_R) can be determined. In finding the time constant $\tau=RC$, R is often the Thevenin equivalent resistance at the terminals of the capacitor; that is, we take out the capacitor C and find $R=R_{Th}$ at its terminals.

Example 4.3.1

The switch in the circuit shown in Figure for example 4.3.1 has been in position x for a long time. At $t=0$, the switch moves instantaneously to position y. Find

(a) $v_C(t)$ for $t\geqslant 0$,

(b) $v_0(t)$ for $t\geqslant 0^+$,

(c) $i_0(t)$ for $t\geqslant 0^+$.

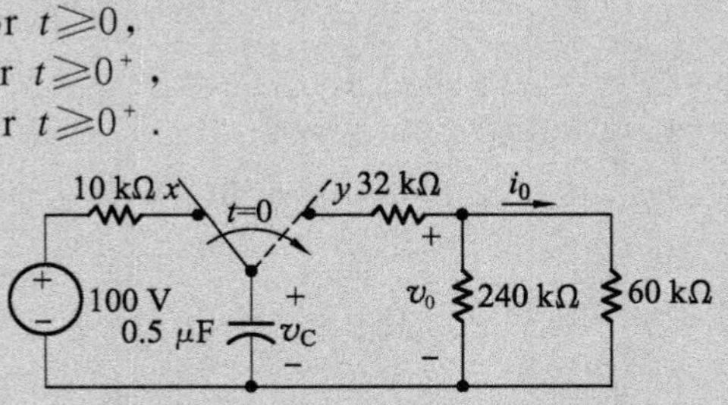

Figure for example 4.3.1

Solution: (a) Because the switch has been in position x for a long time, the 0.5 mF capacitor will charge to 100 V and be positive at the upper terminal. We can replace the resistive network connected to the capacitor at $t=0^+$ with an equivalent resistance of 80 kΩ. Hence the time constant of the circuit is

$$\tau=0.5\times10^{-6}\times(32+240 \parallel 60)\times10^{3}=40 \text{ ms}$$

Then

$$v_C(t)=v_0 e^{\frac{-t}{\tau}}=100e^{-25t}\text{V}, \ t\geqslant0$$

(b) The easiest way to find $v_0(t)$ is to note the resistive circuit forms a voltage divider across the terminals of the capacitor. Thus,

$$v_0(t)=\frac{48}{80}v_C(t)=60e^{-25t} \text{ V}, \ t\geqslant0^+$$

This expression for $v_0(t)$ is valid for $t\geqslant0^+$ because $v_0(0^-)$ is zero. Thus, we have an instantaneous change in the voltage across the 240 kΩ resistor.

(c) We find the current $i_0(t)$ from Ohm's law.

$$i_0(t)=\frac{v_0(t)}{60\times10^3}=e^{-25t} \text{ mA}, \ t\geqslant0^+$$

4.4 The natural response of an RL circuit

Consider the series connection of a resistor and an inductor, as shown in Figure 4.4.1.

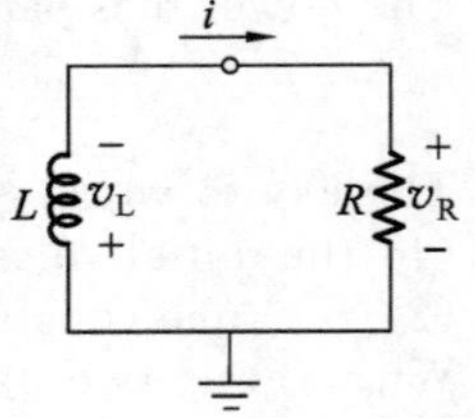

Figure 4.4.1 A source-free of RL circuit

4.4.1 Deriving the expression for the current

Our goal is to determine the circuit response, which we will assume to be the current $i(t)$ through the inductor. We select the inductor current as the response in order to take advantage of the idea that the inductor current cannot change instantaneously. At $t=0$, we assume that the inductor has an initial current I_0, or

$$i(0)=I_0 \tag{4.4.1}$$

Apply KVL around the loop in Figure 4.4.1,

$$v_R+v_L=0 \tag{4.4.2}$$

But $v_L=L\frac{di}{dt}$ and $v_R=iR$. Thus,

$$L\frac{di}{dt}+Ri=0$$

or

$$\frac{di}{dt}+\frac{R}{L}i=0 \tag{4.4.3}$$

Rearranging terms and integrating gives

$$\int_{I_0}^{i(t)}\frac{di}{i}=-\int_0^t dt$$

or

$$\ln\frac{i(t)}{I_0}=-\frac{Rt}{L} \tag{4.4.4}$$

Taking the powers of e, we have

$$i(t)=I_0 e^{\frac{-Rt}{L}} \tag{4.4.5}$$

4.4.2 The significance of the time constant

This shows that the natural response of the RL circuit is an exponential decay of the initial current. The current response is shown in Figure 4.4.2.

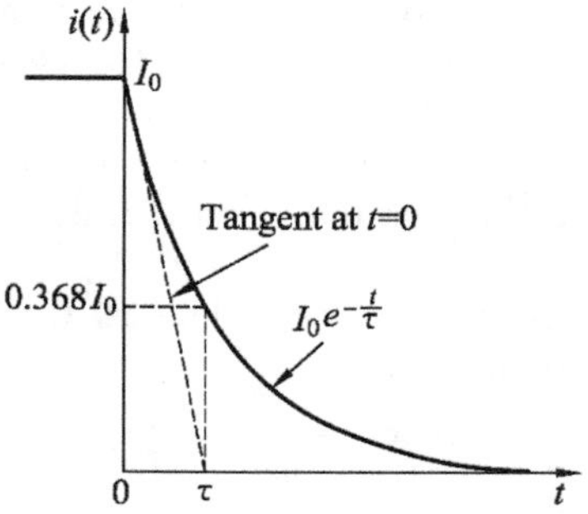

Figure 4.4.2 The current response of the RL circuit

It is evident from Equation (4.4.5) that the time constant for the RL circuit is

$$\tau=\frac{L}{R} \tag{4.4.6}$$

Thus, Equation (4.4.5) may be written as

$$i(t)=I_0 e^{\frac{-t}{\tau}} \tag{4.4.7}$$

With the current in Equation (4.4.7), we can find the voltage across the resistor as

$$v_R(t)=iR=I_0 R e^{\frac{-t}{\tau}} \tag{4.4.8}$$

The power dissipated in the resistor is

$$p(t)=vi_R=I_0^2 R e^{\frac{-2t}{\tau}} \tag{4.4.9}$$

The key to working with a source-free RC circuit is finding:

(1) the initial voltage $i(0)=I_0$ across the capacitor;

(2) the time constant τ.

With these two items, we obtain the response as the capacitor voltage $i_L(t)=i(t)=i(0)e^{\frac{-t}{\tau}}$. Once the inductor current is first obtained, other variables (inductor voltage v_L, resistor voltage v_R, and resistor current i_R) can be determined. In finding the time constant $\tau=\frac{L}{R}$, R is often the Thevenin equivalent resistance at the terminals of the inductor; that is, we take out the inductor L, and find $R=R_{Th}$ at its terminals.

Example 4.4.1

The switch in the circuit shown in Figure for example 4.4.1 has been closed for a long time before it is opened. Find (a) $i_L(t)$ for $t\geqslant 0$, (b) $i_0(t)$ for $t\geqslant 0^+$, (c) $v_0(t)$ for $t\geqslant 0^+$.

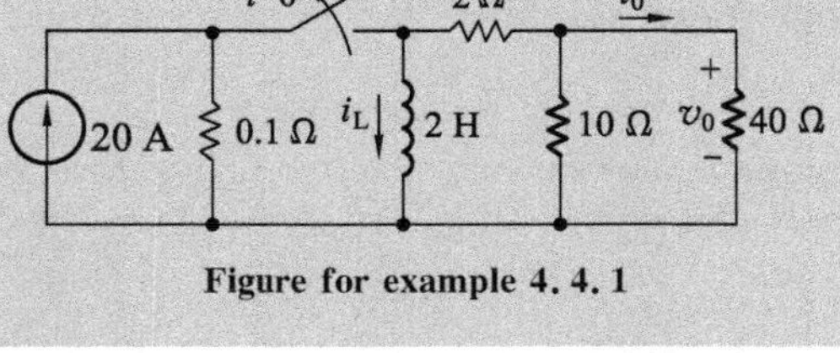

Figure for example 4.4.1

Solution: (a) The switch has been closed for a long time prior to $t=0$, so we know the voltage across the inductor must be zero at $t=0^-$. Therefore the initial current in the inductor is 20 A at $t=0^-$. Hence, $i_L(0^+)$ also is 20 A, because an instantaneous change in the current cannot occur in an inductor. We replace the resistive circuit connected to the terminals of the inductor with a single resistor of 10 Ω:

$$R_{eq}=2+(40 \parallel 10)=10\ \Omega$$

The time constant of the circuit is

$$\tau=\frac{L}{R_{eq}}=\frac{2}{10}=0.2\ \text{s}$$

So

$$i_L(t)=20e^{-5t}\ \text{A},\ t \geqslant 0$$

(b) We find the current in the 40 Ω resistor most easily by using current division; that is

$$i_0=-i_L\frac{10}{10+40}$$

Note that this expression is valid for $t \geqslant 0^+$, because $i_0=0$ at $t=0^-$. The inductor behaves as a short circuit prior to the switch being opened, producing an instantaneous change in the current i_0. Then

$$i_0(t)=-4e^{-5t}\ \text{A},\ t \geqslant 0^+$$

(c) We find the voltage v_0 by direct application of Ohm's law:

$$v_0(t)=40i_0=-160e^{-5t}\ \text{V},\ t \geqslant 0^+$$

4.5 The step response of an RC circuit

When the DC source of an RC circuit is suddenly applied, the voltage or current source can be modeled as a step function, and the response is known as a step response.

Consider the RC circuit in Figure 4.5.1(a), which can be replaced by the circuit in V_s Figure 4.5.1(b), where V_s is a constant DC voltage source. Again, we select the capacitor voltage as the circuit response to be determined.

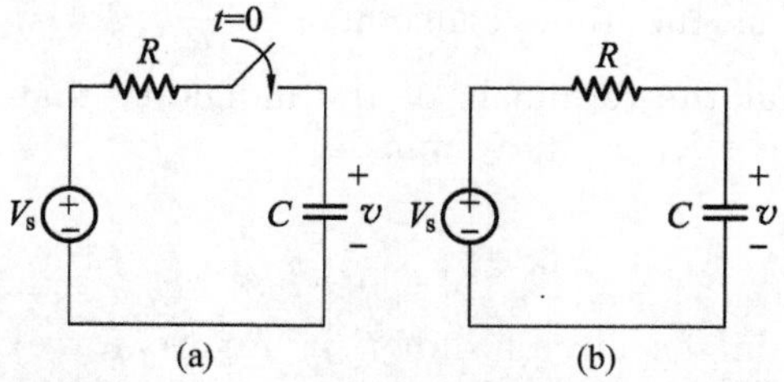

Figure 4.5.1 An RC circuit with voltage step input

We assume an initial voltage V_0 across the capacitor, although this is not necessary for the step response. Since the voltage of a capacitor cannot change instantaneously,

$$v(0^-)=v(0^+)=V_0 \tag{4.5.1}$$

where $v(0^-)$ is the voltage across the capacitor just before switching and $v(0^+)$ is its voltage immediately after switching. Applying KCL, we have

$$C\frac{dv}{dt}+\frac{v-V_s(t)}{R}=0$$

or

$$\frac{dv}{dt}+\frac{v}{RC}=\frac{V_s(t)}{RC} \tag{4.5.2}$$

where v is the voltage across the capacitor. Rearranging terms gives

$$\frac{dv}{dt}=-\frac{v-V_s}{RC}$$

or

$$\frac{dv}{v-V_s}=-\frac{dt}{RC} \tag{4.5.3}$$

Integrate both sides and introducing the initial conditions, and we have

$$\ln(v-V_s)\Big|_{V_0}^{v(t)}=-\frac{t}{RC}\Big|_0^t$$

$$\ln(v(t)-V_s)-\ln(V_0-V_s)=-\frac{t}{RC}+0$$

or

$$\ln\frac{(v-V_s)}{(V_0-V_s)}=-\frac{t}{RC} \tag{4.5.4}$$

Taking the exponential of both sides,

$$\frac{v-V_s}{V_0-V_s}=e^{-\frac{t}{\tau}}, \tau=RC$$

or

$$v(t)=V_s+(V_0-V_s)e^{-\frac{t}{\tau}} \tag{4.5.5}$$

Thus,

$$v(t)=\begin{cases}V_0 & t<0\\ V_s+(V_0-V_s)e^{-\frac{t}{\tau}} & t>0\end{cases} \tag{4.5.6}$$

This is known as the complete response (or total response) of the RC circuit to a sudden application of a DC voltage source, assuming the capacitor is initially charged. The reason for the term "complete" will become evident a little later. Assuming that $V_s>V_0$, a plot of $v(t)$ is shown in Figure 4.5.2.

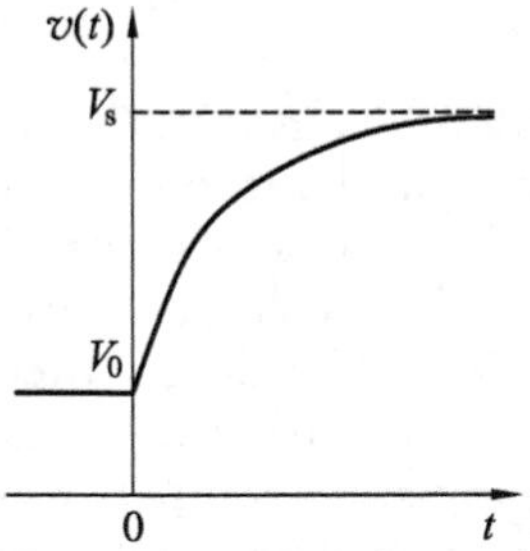

Figure 4.5.2 Response of an RC circuit with intially charged capacitor

If we assume that the capacitor is uncharged initially, we set $V_0=0$ in Equation (4.5.6) so that

$$v(t)=\begin{cases}0 & t<0\\ V_s(1-e^{-\frac{t}{\tau}}) & t>0\end{cases} \tag{4.5.7}$$

This is the complete step response of the RC circuit when the capacitor is initially uncharged. The current through the capacitor is obtained from Equation

(4. 5. 7) using $i(t)=C\dfrac{dv}{dt}$. We get

$$i(t)=C\frac{dv}{dt}=\frac{C}{\tau}V_s e^{-\frac{t}{\tau}},\ \tau=RC,\ t>0$$

or

$$i(t)=\frac{V_s}{R}e^{-\frac{t}{\tau}} \tag{4.5.8}$$

The plot of $u(t)$ and $i(t)$ are shown in Figure 4. 5. 3.

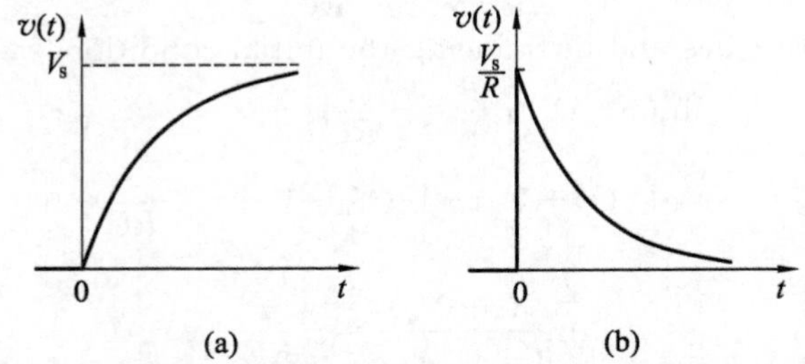

Figure 4. 5. 3 The plots of capacitor voltage $v(t)$ and capacitor current $i(t)$

Rather than going through the derivations above, there is a systematic approach, or rather, a short cut method—for finding the step response of an RC or RL circuit. Let us reexamine Equation (4. 5. 5), which is more general than Equation (4. 5. 7). It is evident that $v(t)$ has two components.

Classically, there are two ways of decomposing this into two components. The first is to break it into a "natural response and a forced response" and the second is to break it into a "transient response and a steady-state response". Starting with the natural response and forced response, we write the total or complete response as

Complete Response = Natural Response + Forced Response

or

$$v=v_n+v_f \tag{4.5.9}$$

where

$$v_n=V_0 e^{-\frac{t}{\tau}},\ v_f=V_s(1-e^{-\frac{t}{\tau}})$$

We are familiar with the natural response v_n of the circuit, as discussed in Section 4. 2. v_f is known as the forced response because it is produced by the circuit when an external "force" (a voltage source in this case) is applied. It represents what the circuit is forced to do by the input excitation. The natural response eventually dies out along with the transient component of the forced response, leaving only the steady-state component of the forced response.

Another way of looking at the complete response is to break into two components—one temporary and the other permanent, i. e. ,

Complete Response = Transient Response + Steady-state Response

or

$$v=v_t+v_{ss} \tag{4.5.10}$$

where

$$v_t=(V_0-V_s)e^{-\frac{t}{\tau}},\ v_{ss}=V_s$$

The transient response v_t is temporary; it is the portion of the complete response that decays to zero as time approaches infinity. Thus,

the transient response is the circuit's temporary response that will die out with time.

The steady-state response v_{ss} is the portion of the complete response that remains after the transient response has die out. Thus,

the steady-state response is the behavior of the circuit a long time after an external excitation is applied.

The first decomposition of the complete response is in terms of the source of the response, while the second decomposition is in terms of the permanency of the response. Under certain conditions, the natural response and transient response are the same. The same can be said about the force response and steady-state response.

In whichever way we look at it, the complete response in Equation (4.5.5) may be written as

$$v(t)=v(\infty)+[v(0)-v(\infty)]e^{-\frac{t}{\tau}} \tag{4.5.11}$$

where $v(0)$ is the initial voltage at $t=0^+$ and $v(\infty)$ is the final or steady-state value. Thus, to find the step response of an RC circuit requires three things:

(1) the initial capacitor voltage $v(0)$.

(2) the final capacitor voltage $v(\infty)$.

(3) the time constant τ.

We obtain item 1 from the given circuit for $t<0$ and items 2 and 3 from the circuit for $t>0$. Once these items are determined, we obtain the response using Equation (4.5.11). This technique equally applies to RL circuits, as we shall see in the next section.

Note that if the switch changes position at time $t=t_0$ instead of at $t=0$, there is a time delay in the response so that Equation (4.5.11) becomes

$$v(t)=v(\infty)+[v(t_0)-v(\infty)]e^{-\frac{(t-t_0)}{\tau}} \tag{4.5.12}$$

where $v(t_0)$ is the initial value at $t=t_0^+$. Keep in mind that Equation (4.5.11) or Equation (4.5.12) applies only to step response, that is, when the input excitation is constant.

Example 4.5.1

The switch in Figure for example 4.5.1 has been in position A for a long time. At $t=0$, the switch moves to B. Determine $v(t)$ *for* $t>0$ *and calculate its value at* $t=1$ *s and* 4 *s*.

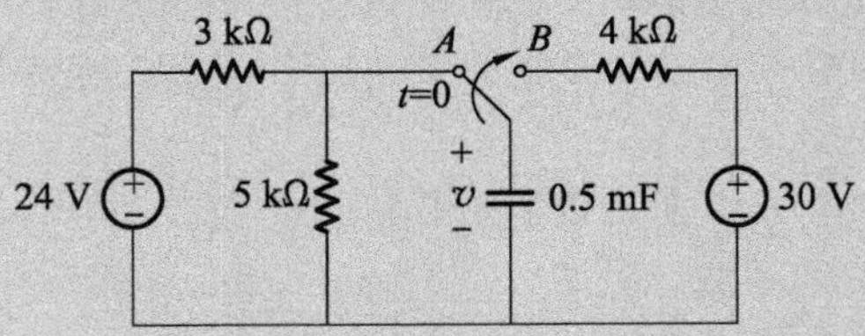

Figure for example 4.5.1

Solution: For $t<0$, the switch is at position A. The capacitor acts like an open circuit to DC, but v is the same as the voltage across the 5 kΩ resistor. Hence, the voltage across the capacitor just before $t=0$ is obtained by voltage division as

$$v(0^-)=\frac{5}{5+3}\times 24=15\ \text{V}$$

Use the fact that the capacitor voltage cannot change instantaneously,

$$v(0^-)=v(0)=v(0^+)=15\ \text{V}$$

For $t>0$, the switch is in position B. The Thevenin resistance connected to the capacitor is

$$R=4\ \text{k}\Omega$$

And the time constant is

$$\tau=RC=4\times 10^3\times 0.5\times 10^{-3}=2\ \text{s}$$

Since the capacitor acts like an open circuit to DC at a steady state,

$$v(\infty)=30\ \text{V}$$

Thus,

$$\begin{aligned} v(t)&=v(\infty)+[v(t_0)-v(\infty)]\text{e}^{-\frac{t}{\tau}} \\ &=30+(15-30)\text{e}^{\frac{-t}{2}} \\ &=30-15\text{e}^{\frac{-t}{2}}\ \text{V} \end{aligned}$$

At $t=1$ s, $v(t)=30-15\text{e}^{-0.5}=20.902$ V

At $t=4$ s, $v(t)=30-15\text{e}^{-2}=27.97$ V

Example 4.5.2

In Figure for example 4.5.2, the switch has been closed for a long time and is opened at $t=0$. Find i and v for all time.

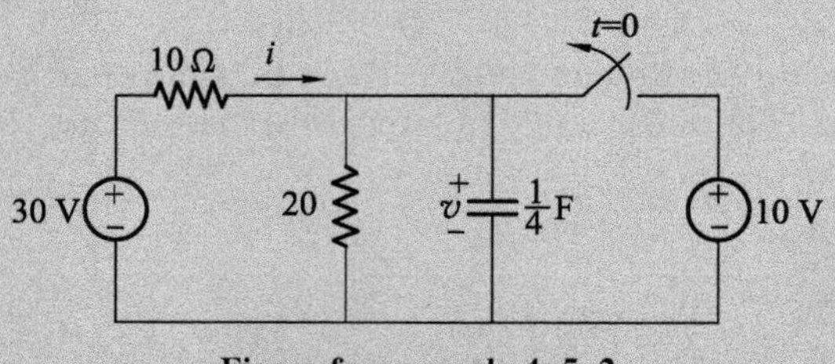

Figure for example 4.5.2

Solution: The resistor current i can be discontinuous at $t=0$, while the capacitor voltage v cannot. Hence, it is always better to find v and obtain i from v.

$$v(0^+)=v(0^-)=10\ \text{V}$$

$$\tau=RC=(10\parallel 20)\times\frac{1}{4}=\frac{5}{3}\ \text{s}$$

$$v(\infty)=\frac{20}{20+10}\times 30=20\ \text{V}$$

Thus,

$$\begin{aligned} v(t) &= v(\infty) + [v(t_0) - v(\infty)]e^{-\frac{t}{\tau}} \\ &= 20 + (10 - 20)e^{-0.6t} \\ &= 20 - 10e^{-0.6t} \text{ V} \end{aligned}$$

To obtain i, we notice from Figure for example 4.5.2 that i is the sum of the currents through the 20 Ω and the capacitor, which is

$$\begin{aligned} i &= \frac{v(t)}{20} + C\frac{dv}{dt} \\ &= 1 - 0.5e^{-0.6t} + 0.25 \times (-0.6)(-10)e^{-0.6t} \\ &= 1 + e^{-0.6t} \text{ A} \end{aligned}$$

Hence,

$$v = \begin{cases} 10 \text{ V} & t < 0 \\ 20 - 10e^{-0.6t} \text{ V} & t \geq 0 \end{cases}$$

$$i = \begin{cases} -1 \text{ A} & t < 0 \\ 1 + e^{-0.6t} \text{ A} & t \geq 0 \end{cases}$$

4.6 The step response of an RL circuit

Consider the RL circuit in Figure 4.6.1 (a), which may be replaced by the circuit in Figure 4.6.1 (b). Again, our goal is to find the inductor current i as the circuit response. Rather than apply Kirchoff's laws, we will use the simple technique in Equation (4.5.9) through Equation (4.5.12). Let the response be the sum of the natural current and the forced current.

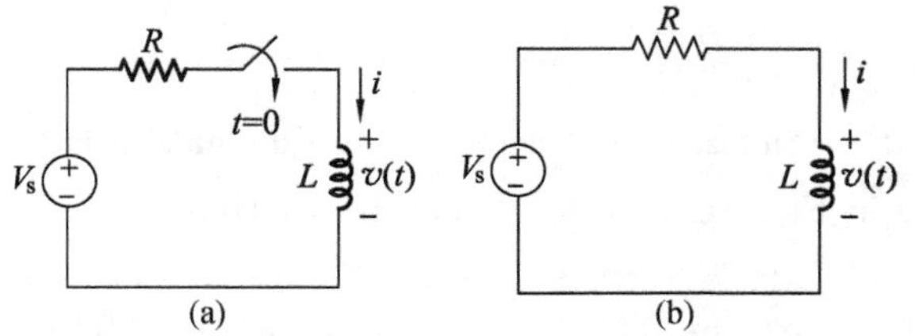

Figure 4.6.1 An RL circuit with a step input voltage

$$i = i_n + i_f \tag{4.6.1}$$

We know that the natural response is always a decaying exponential, which is

$$i_n = Ae^{\frac{-t}{\tau}}, \quad \tau = \frac{L}{R} \tag{4.6.2}$$

where A is a constant to be determined.

The forced response is the value of the current a long time after the switch in Figure 4.6.1 (a) is closed. We know that the natural response essentially dies out after five time constants. At that time, the inductor V_s becomes a short-circuit, and the voltage across it is zero. The entire source voltage v_s appears across R. Thus, the forced response is

$$i_f = \frac{V_s}{R} \tag{4.6.3}$$

Substituting Equation (4.6.2) and Equation (4.6.3) into Equation (4.6.1) gives

$$i = Ae^{\frac{-t}{\tau}} + \frac{V_s}{R} \tag{4.6.4}$$

We now determine the constant A from the initial value of i. Let I_0 be the initial current through the inductor, which may come from a source other than V_s, since the current through the inductor cannot change instantaneously.

$$i(0^+)=i(0^-)=I_0 \tag{4.6.5}$$

Thus, at $t=0$, Equation (4.6.4) becomes

$$I_0=A+\frac{V_s}{R}$$

From this, we obtain A as

$$A=I_0-\frac{V_s}{R}$$

Substituting for A in Equation (4.6.4), we get

$$i(t)=\frac{V_s}{R}+\left(I_0-\frac{V_s}{R}\right)e^{\frac{-t}{\tau}} \tag{4.6.6}$$

This is the complete response of the RL circuit. It is illustrated in Figure 4.6.2.

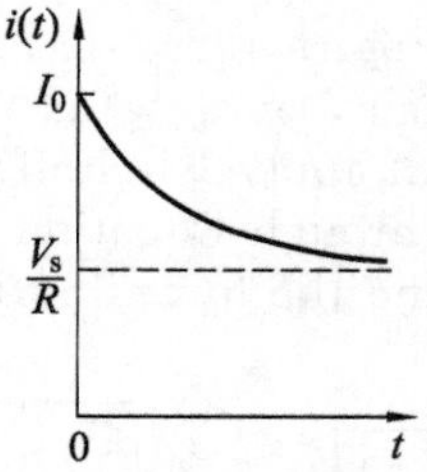

Figure 4.6.2 Total responses of the RL circuit with initial inductor current I_0

The response in Equation (4.6.6) may be written as

$$i(t)=i(\infty)+[i(t_0)-i(\infty)]e^{-\frac{(t-t_0)}{\tau}} \tag{4.6.7}$$

where $i(0)$ and $i(\infty)$ are the initial and final values of i. Thus, to find the step response of an RL circuit requires three things.

(1) the initial inductor current $i(0)$.

(2) the final inductor current $i(\infty)$.

(3) the time constant τ.

We obtain item 1 from the given circuit for $t<0$ and items 2 and 3 from the circuit for $t>0$. Once these items are determined, we obtain the response using Equation (4.6.7). Keep in mind that this technique applies only to step response.

Again if the switchs take place at time $t=t_0$ instead of at $t=0$, Equation (4.6.7) becomes

$$i(t)=i(\infty)+[i(t_0)-i(\infty)]e^{-\frac{(t-t_0)}{\tau}} \tag{4.6.8}$$

If $I_0=0$, then

$$i(t)=\begin{cases}0 & t<0\\ \frac{V_s}{R}(1-e^{\frac{-t}{\tau}}) & t>0\end{cases} \tag{4.6.9}$$

This is the step response of the RL circuit with no initial inductor current. The voltage across the inductor is obtained from Equation (4.6.9) using $v=$

$L\frac{di}{dt}$. We get

$$v(t)=L\frac{di}{dt}=V_s\frac{L}{\pi R}e^{\frac{-t}{\tau}},\ \tau=\frac{L}{R},\ t>0 \tag{4.6.10}$$

Figure 4.6.3 shows the step response in Equations (4.6.9) and (4.6.10).

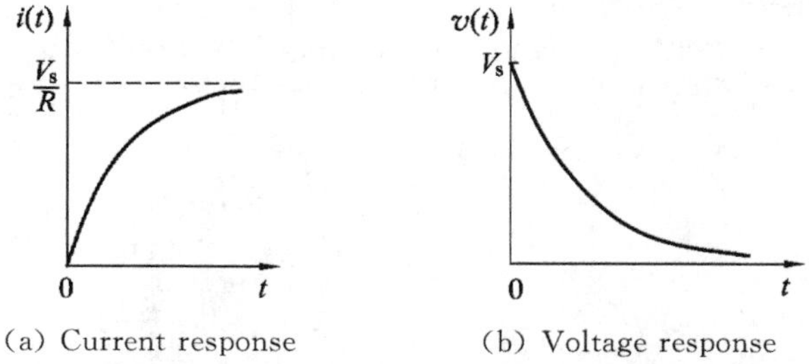

Figure 4.6.3 Step response of an RL circuit with no initial inductor current

Example 4.6.1

Find $i(t)$ in the circuit in Figure for example 4.6.1 for $t>0$. Assume that the switch has been closed for a long time.

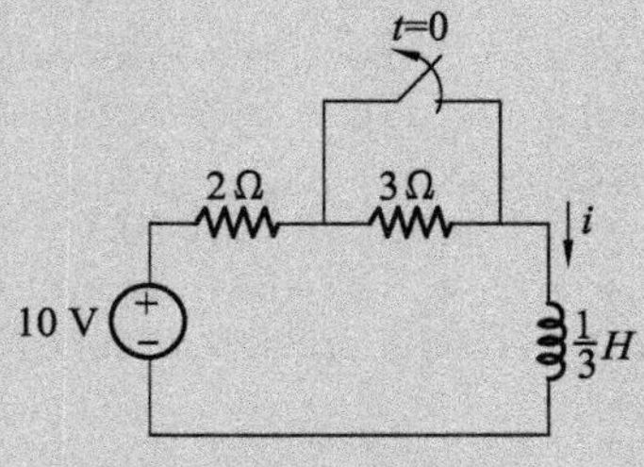

Figure for example 4.6.1

Solution: When, the 3 Ω resistor is short-circuited, and the inductor acts like a short-circuit. The current through the inductor at $t=0^-$ is

$$i(0^-)=\frac{10}{2}=5\text{ A}$$

Since the inductor current cannot change instantaneously,

$$i(0)=i(0^+)=i(0^-)=5\text{ A}$$

When $t>0$, the switch is open. The 2 Ω and 3 Ω resistors are in series, so that

$$i(\infty)=\frac{10}{2+3}=2\text{ A}$$

The Thevenin resistance across the inductor terminals is

$$R=2+3=5\ \Omega$$

For the time constant,

$$\tau=\frac{L}{R}=\frac{\frac{1}{3}}{5}=\frac{1}{15}\text{ s}$$

Thus,

$$i(t)=i(\infty)+[i(0)-i(\infty)]e^{\frac{-t}{\tau}}$$
$$=2+(5-2)e^{-15t}=2+3e^{-15t}\ \text{A}$$

Example 4.6.2

At $t=0$, switch 1 in Figure for example 4.6.2 is closed, and switch 2 is closed 4 s later. Find $i(t)$ for $t>0$. Calculate i for $t=2$ s and $t=5$ s.

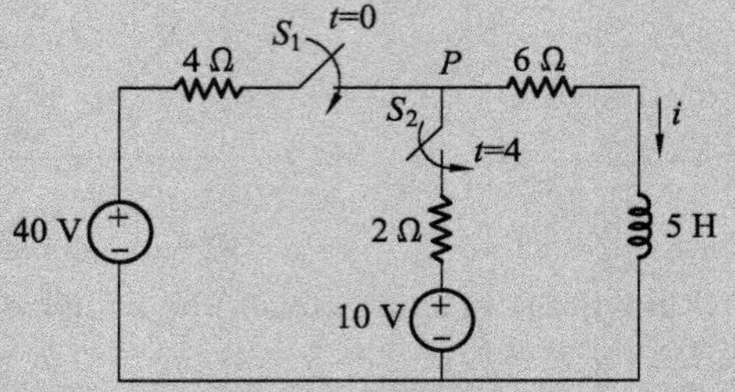

Figure for example 4.6.2

Solution: We need to consider the three intervals $t\leqslant 0$, $0\leqslant t\leqslant 4$ s, $t\geqslant 4$ s separately. For $t<0$, switches S_1 and S_2 are open, so that. Since the inductor current cannot change instantly,

$$i(0^-)=i(0)=i(0^+)=0$$

For $0\leqslant t\leqslant 4$ s, S_1 is closed so that the 4 Ω and 6 Ω resistors are in series. (Remember, at this time, S_2 is still open.) Hence, assume for now that S_1 is closed forever.

$$i(\infty)=\frac{40}{4+6}=4\ \text{A},\ R=4+6=10\ \Omega$$

$$\tau=\frac{L}{R}=\frac{5}{10}=\frac{1}{2}\ \text{s}$$

Thus,

$$i(t)=i(\infty)+[i(0)-i(\infty)]e^{\frac{-t}{\tau}}$$
$$=4+(0-4)e^{-2t}=4-4e^{-2t}\ \text{A},\ 0\leqslant t\leqslant 4\ \text{s}$$

For $t\geqslant 4$ s, S_2 is closed; the 10 Ω voltage source is connected, and the circuit changes. This sudden change does not affect the inductor current because

$$5\tau=2.5\ \text{s}$$

For $t\geqslant 4$ s, the circuit already gets the new steady state after $(3\sim5)\tau$. Thus, the initial current is

$$i(4)=i(4^-)=4\ \text{A}$$

To find $i(\infty)$, let v be the voltage at node P in Figure for example 4.6.2 (Remember we consider inductor as a short circuit when the circuit is in the steady state). Using KCL,

$$\frac{40-v}{4}+\frac{10-v}{2}=\frac{v}{6}\Rightarrow v=\frac{180}{11}\ \text{V}$$

$$i(\infty)=\frac{v}{6}=\frac{30}{11}=2.727\ \text{A}$$

$$R=4\parallel 2+6=\frac{22}{3}\ \Omega,\tau=\frac{L}{R}=\frac{15}{22}\ \text{s}$$

Hence,

$$\begin{aligned}i(t)&=i(\infty)+[i(0)-i(\infty)]e^{\frac{-(t-4)}{\tau}}\\&=2.727+(4-2.727)e^{\frac{-(t-4)}{\tau}}\\&=2.727+1.273e^{-1.4667t}\ \text{A}\qquad t\geqslant 4\ \text{s}\end{aligned}$$

Putting all this together,

$$i(t)=\begin{cases}0 & t\leqslant 0\\ 4(1-e^{-2t}) & 0\leqslant t\leqslant 4\ \text{s}\\ 2.727+1.273e^{-1.4667(t-4)} & t\geqslant 4\ \text{s}\end{cases}$$

At $t=2$ s, $i(2)=4(1-e^{-4})=3.93$ A.

At $t=5$ s, $i(5)=2.727+1.273e^{-1.4667}=3.02$ A.

4.7 A general solution for step and natural responses

The general approach to finding either the natural response or the step response of the first-order RL and RC circuits shown in Figure 4.7.1 is based on the different equations having the same form [compared with Equation (4.5.2)]. To generalize the solution of these four possible circuits, we let $x(t)$ represent the unknown quantity, giving $x(t)$ four possible values. It can represent the current or voltage at the terminals of a capacitor. From Equations (4.3.4), (4.4.3) and (4.5.2), we know that the differential equation describing any one of the four circuits in Figure 4.7.1 takes the form

$$\frac{dx}{dt}+\frac{x}{\tau}=K \tag{4.7.1}$$

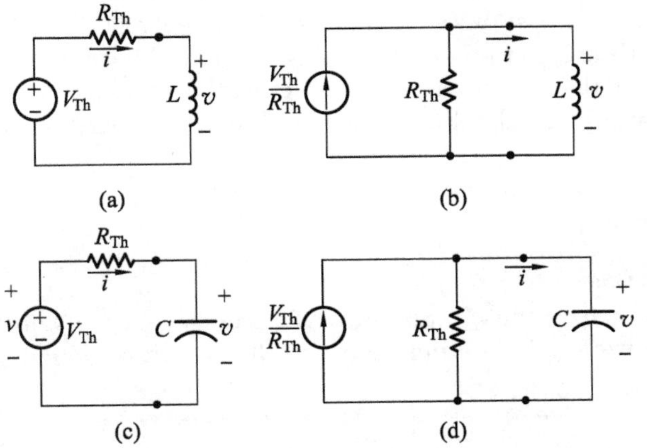

Figure 4.7.1 Four possible first-order circuits

In Equation(4.7.1), the value of the constant K can be zero. Because the sources in the circuit are constant voltage and/or currents, the final value of x will be constant; that is, the final value must satisfy Equation (4.7.1), and, when x reaches its final value, the derivative dx/dt must be zero. Hence,

$$x_f=K\tau \tag{4.7.2}$$

where x_f represents the final value of the variable.

We solve Equation (4.7.1) by separating the variables, beginning by solving for the first derivative:

$$\frac{dx}{dt}=\frac{-x}{\tau}+K=\frac{-(x-K\tau)}{\tau}=\frac{-(x-x_f)}{\tau} \tag{4.7.3}$$

In Equation (4.7.3), we used Equation (4.7.2) to substitute x_f for $K\tau$. We now multiply both sides of Equation (4.7.3) by dt and divide by $x-x_f$ to obtain

$$\frac{dx}{x-x_f}=\frac{-1}{\tau}dt \tag{4.7.4}$$

Next, we integrate Equation (4.7.4) to obtain as general a solution as possible. We use time as the lower limit and t as the upper limit. Time t_0 corresponds to the time of the switching or other change. Previously we assumed that t_0, but this change allows the switching to take place at any time. Using u and v as symbols of integration, we get

$$\int_{x(t_0)}^{x(t)}\frac{du}{u-x_f}=\frac{1}{\tau}\int_{t_0}^{t}dv \tag{4.7.5}$$

Carrying out the integration called for in Equation (4.7.5) gives

$$x(t)=x_f+[x(t_0)-x_f]e^{\frac{-(t-t_0)}{\tau}} \tag{4.7.6}$$

In many cases, the time of switching that is, t_0 is zero.

When computing the step and natural response of circuits, it may help to follow these steps:

(1) Identify the variable of interest for the circuit. For RC circuits, it is the most convenient to choose the capacitive voltage; for RL circuits, it is the best to choose the inductive current.

(2) Determine the initial value of the variable t_0, which is its value at t_0. Note that if you choose capacitive voltage or inductive current as your variable of interest, it is not necessary to distinguish between $t=t_0^-$ and $t=t_0^+$. This is because they both are continuous variables. If you choose another variable, you need to remember that its initial value is defined at $t=t_0^+$.

(3) Calculate the final value of the variable of the variable, which is its value as $t\to\infty$.

(4) Calculate the time constant for the circuit.

Example 4.7.1

The switch in the circuit shown in Figure A for example 4.7.1 has been in position a for a long time. At the switch is moved to position b.

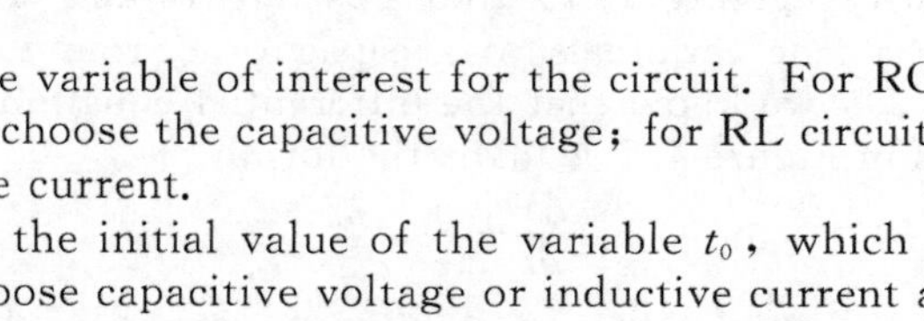

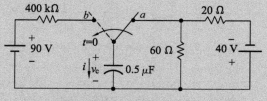

Figure A for example 4.7.1

(a) What is the initial value of v_C?
(b) What is the final value of v_C?
(c) What is the time constant of the circuit when the switch is in position b?
(d) What is the expression for $v_C(t)$ when $t \geqslant 0$?
(e) What is the expression for $i(t)$ when $t \geqslant 0$?
(f) How long after the switch is in position b does the capacitor voltage equal zero?
(g) Plot $v_C(t)$ and $i(t)$ versus t.

Solution: (a) The switch has been in position a for a long time, so the capacitor looks like an open-circuit. Therefore the voltage across the capacitor is the voltage across the 60 Ω resistor. From the voltage-divider rule, the voltage across the 60 Ω resistor is

$$v_{60\ \Omega}=\frac{60}{60+20}\times 40=30\ \text{V}$$

As the reference for v_C is positive at the upper terminal of the capacitor, we have

$$v_C(0)=-30\ \text{V}$$

(b) After the switch has been in position b for a long time, the capacitor will look like an open-circuit in terms of the 90 V source. Thus the final value of the capacitor across is 90 V.

(c) The time constant is

$$\tau=RC=(400\times 10^3)(0.5\times 10^{-6})=0.2\ \text{s}$$

(d) Substituting the appropriate values for v_f, $v(0)$ and t into Equation (4.7.6) yields

$$v_C(t)=90+(-30-90)e^{-5t}$$
$$=90-120e^{-5t}\ \text{V}, t\geqslant 0$$

Here the value of τ doesn't change. Thus we only need to find the initial and final values for the current in the capacitor. When obtaining the initial value, we must get the value of $i(0^+)$, because the current in the capacitor can change instantaneously. This current is equal to the current in the resistor, which from Ohm's law is

$$i(0^+)=\frac{90-(-30)}{400\times 10^3}=300\ \mu\text{A}$$

Note that when applying Ohm's law we recognize that the capacitor voltage cannot change instantaneously. The final value of $i(t)=0$, so

$$i(t)=0+(300-0)e^{-5t}$$
$$=300e^{-5t}\ \mu\text{A}, t\geqslant 0^+$$

(e) We could have obtained this solution by differentiating the solution in (d) and multiplying by the capacitance. You may want to do so for yourself. Note that this alternative approach to finding $i(t)$ also predicts the discontinuity at $t=0$.

(f) To find how long the switch must be in position b before the capacitor voltage becomes zero, we solve the equation derived in (d) for the time when $v_C(t)=0$

$$120e^{-5t}=90$$

So

$$t=\frac{1}{5}\ln\left(\frac{4}{3}\right)=57.54\ \text{ms}$$

Note that when $v_C=0$, $i=225\ \mu A$ and the voltage drop across the 400 kΩ resistor is 90 V.

(g) Figure B for example 4.7.1 shows the graphs of $v_C(t)$ and $i(t)$ versus t.

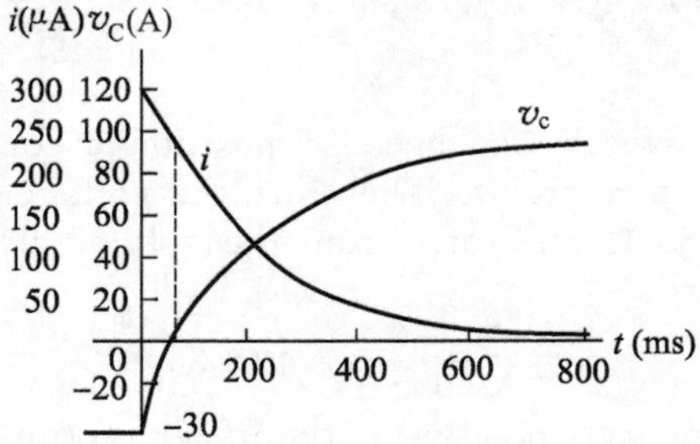

Figure B for example 4.7.1

Example 4.7.2

The switch in the circuit shown in Figure for example 4.7.2 has been open for a long time. At $t=0$ the switch is closed. Find the expression for

(a) $v(t)$ when $t \geqslant 0^+$.

(b) $i(t)$ when $t \geqslant 0$.

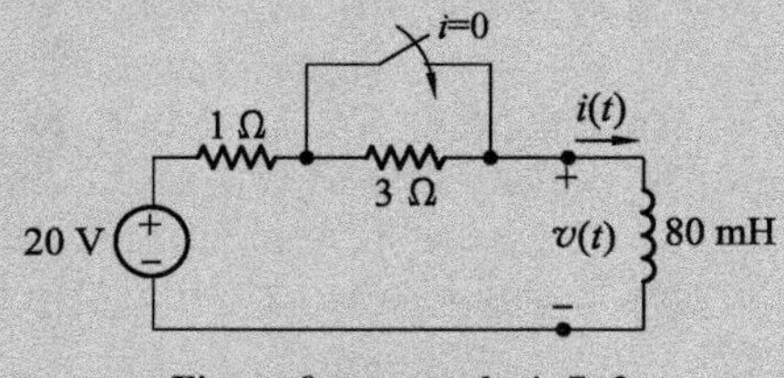

Figure for example 4.7.2

Solution: (a) The switch has been open for a long time, so the initial current in the inductor is

$$i=\frac{20}{1+3}=5\ \text{A}$$

Oriented from top to bottom. Immediately after the switch closes, the current is still 5 A and therefore, the initial voltage across the inductor becomes

$$v(t_0^+)=20-5\times1=15\ \text{V}$$

The final value of the inductor voltage is 0. With the switch closed, the time constant is

$$\tau=\frac{L}{R}=\frac{80}{1}=80\ \text{ms}$$

We use Equation (4.7.6) to write the expression for $v(t)$:

$$v(t)=0+(15-0)e^{\frac{-t}{80\times10^{-3}}}$$
$$=15e^{-12.5t}\ \text{V},\quad t\geqslant0^+$$

(b) We have already noted that the initial value of the inductor current is 5 A. After the switch has been closed for a long time, the inductor current is

$$i=\frac{20}{1}=20\ \text{A}$$

The circuit time constant is 80 ms, so the expression for $i(t)$ is

$$i(t)=20+(5-20)e^{-12.5t}$$
$$=20-15e^{-12.5t}\ \text{A},\quad t\geqslant0$$

We determine that the solutions for $v(t)$ and $i(t)$ agree by noting that

$$v(t)=L\,\frac{di}{dt}=80\times10^{-3}(15\times12.5e^{-12.5t})$$
$$=15e^{-12.5t}\ \text{V},t\geqslant0^+$$

4.8 Simulation of transient circuit using Multisim

4.8.1 Simulation of RC circuit

Place the elements and instruments, and create the circuit as in Figure 4.8.1.

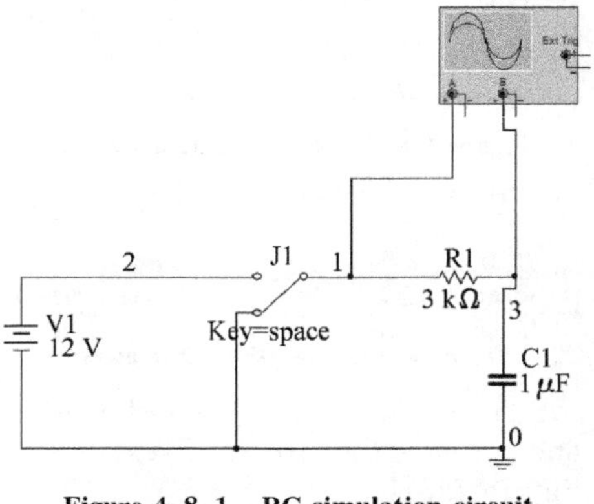

Figure 4.8.1 RC simulation circuit

Double click XSC1, enlarge the panel, and set the parameters as shown in Figure 4.8.2.

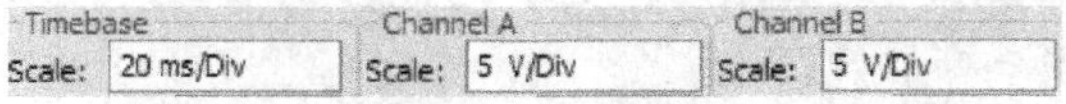

Figure 4.8.2

Connect one side of J1 to DC source, and the other side to GND. Press simulate button, and activate the circuit. Then press button pause, and lock the wave, just like in Figure 4.8.3. Change the value of voltage, and get the different simulation wave.

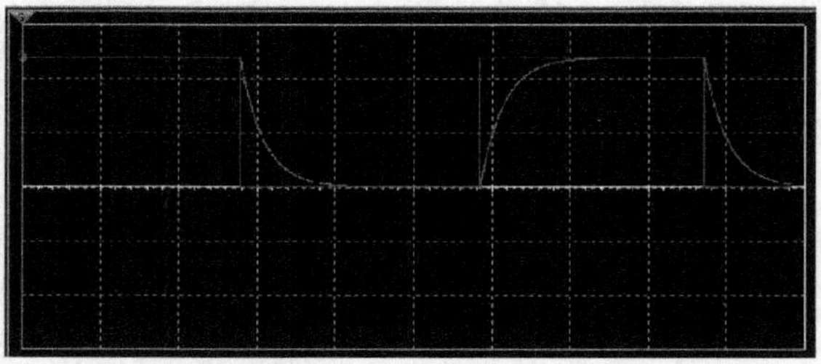

Figure 4.8.3 Wave of simulation

4.8.2 Simulation of RL circuit

Place the elements and instruments, and create the circuit as in Figure 4.8.4.

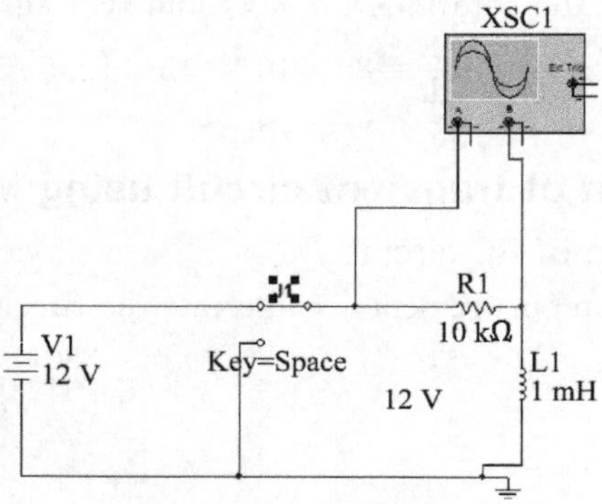

Figure 4.8.4 RL simulation circuit

Double click XSC1, enlarge the panel, and set the parameters as shown in Figure 4.8.5.

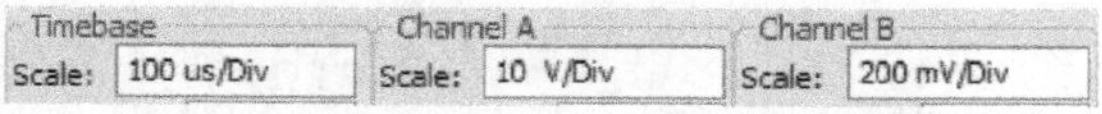

Figure 4.8.5 Set the parameters

Connect one side of J1 to DC source, and the other side to GND. Press simulate button, and activate the circuit. Press button pause, and lock the wave, just like in Figure 4.8.6.

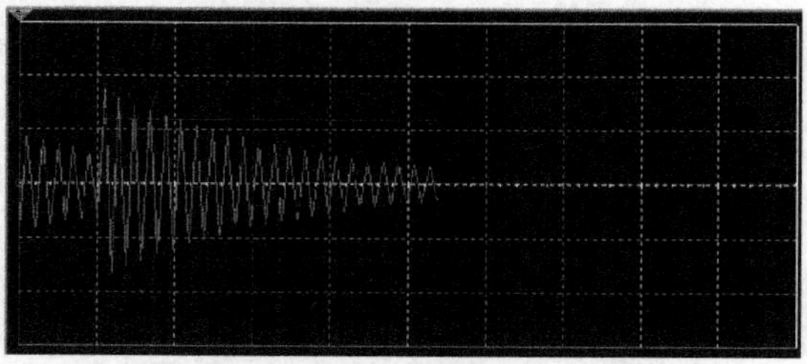

Figure 4.8.6 Wave of simulation

4.8.3 Measure the time constant of an one-order RC circuit

Place the elements and instruments, and create the circuit as in Figure 4.8.7.

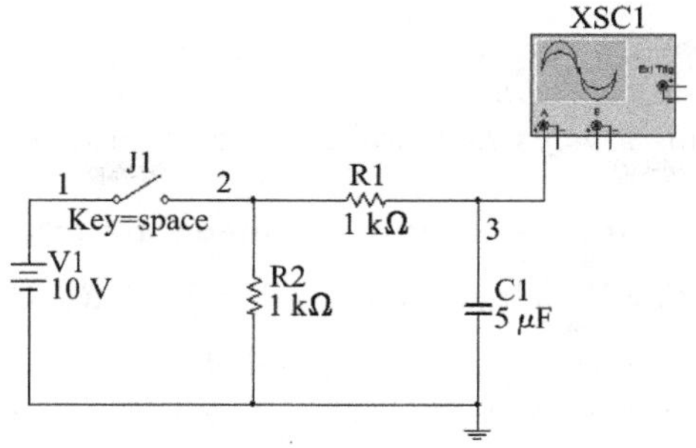

Figure 4.8.7 RC circuit

Double click XSC1, enlarge the panel, and set the parameters as shown in Figure 4.8.8.

Figure 4.8.8 Set the parameters

Press simulate button [icon], activate the circuit, and get wave shown in Figure 4.8.9(a). Change C1, and get the different wave shown in Figure 4.8.9 (b) clearly.

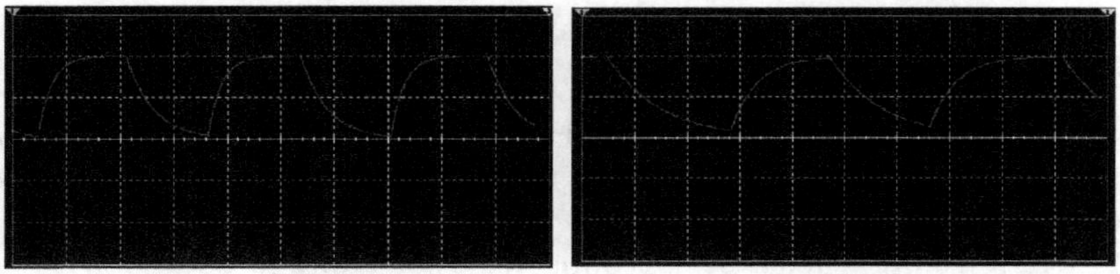

(a) (b)

Figure 4.8.9 Simulation of wave

Summary

1. The analysis in this chapter is applicable to any circuit that can be reduced to an equivalent circuit comprising a resistor and a single energy stored element (inductor and capacitor). Such a circuit is first-order because its behavior is described by a first-order differential equation. When analyzing RC and RL circuits, one must always keep in mind that the capacitor is an open circuit under steady-state DC condition while the inductor is a short circuit under steady-state DC condition.

2. The natural response is obtained when no independent source is present. It has the general form

$$x(t)=x(0)e^{\frac{-t}{\tau}}$$

where x represents current through (or voltage across) a resistor, a capacitor, or an inductor, and $x(0)$ is the initial value of x. Because most practical applliccations always have losses, the natural response is a transient response.

3. The time constant τ is the time required for a response to decay to $1/\tau$ of its initial value. For RC circuits, $\tau=RC$ and for RL circuits, $\tau=\frac{L}{R}$.

4. The steady-state response is the behavior of the circuit after an independent source has been applied for a long time. The transient response is the component of the complete response that dies out with time.

5. The total or complete response consists of the steady-state response and the transient response.

6. The step response is the response of the circuit to a sudden application of a DC current or voltage. Finding the step response of a first-order circuit requires the initial value $x(0^+)$, the final value $x(\infty)$, and the time constant τ. With these three items, we obtain the step response as

$$x(t)=x(\infty)+[x(0^+)-x(\infty)]e^{\frac{-t}{\tau}}$$

Or we may write it as

$$\text{Instantaneous value}=\text{Final}+(\text{Initial-Final})e^{\frac{-t}{\tau}}$$

Problems

4.1 The terminal voltage of a 2 H inductor is $v=10(1-t)$ V. Find the current flowing through it at $t=4$ s and the energy stored in it within $0<t<4$ s. Assume $i(0)=2$ A.

4.2 If a 10 μF capacitor is connected to a voltage source with $v(t)=50\sin 2\,000t$ V, determine the current through the capacitor.

4.3 Determine v_C, i_L, and the energy stored in the capacitor and inductor in the circuit of Figure for problem 4.3 under DC conditions.

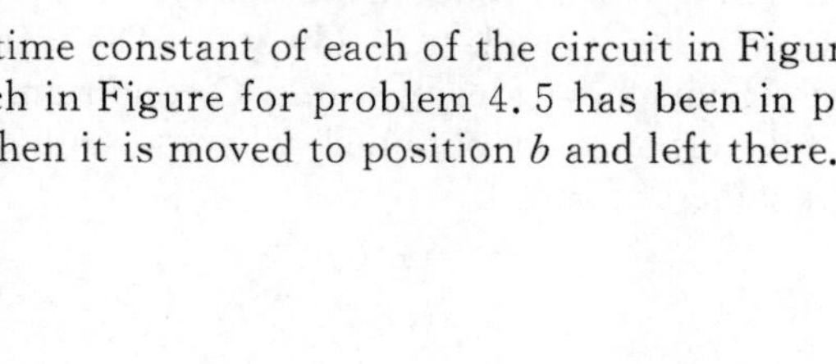

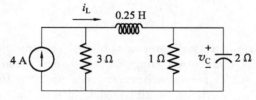

Figure for problem 4.3

4.4 Find the time constant of each of the circuit in Figure for problem 4.4.

4.5 The switch in Figure for problem 4.5 has been in position a for a long time, until $t=4$ s when it is moved to position b and left there. Determine $v(t)$ at $t=10$ s.

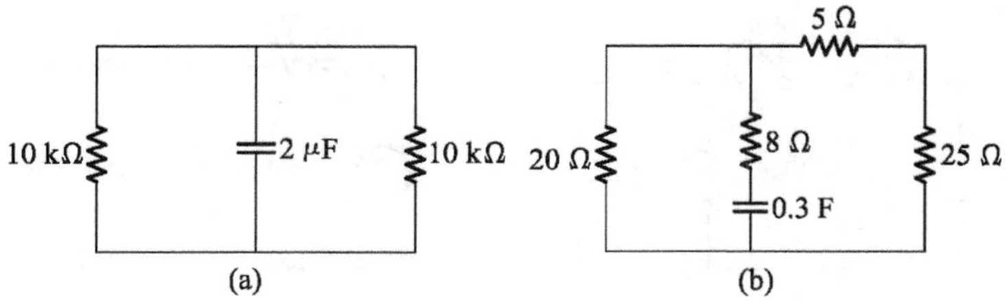

Figure for problem 4. 4

4. 6 In the circuit of Figure for problem, $v(0)=20$ V. Find $v(t)$ for $t>0$.

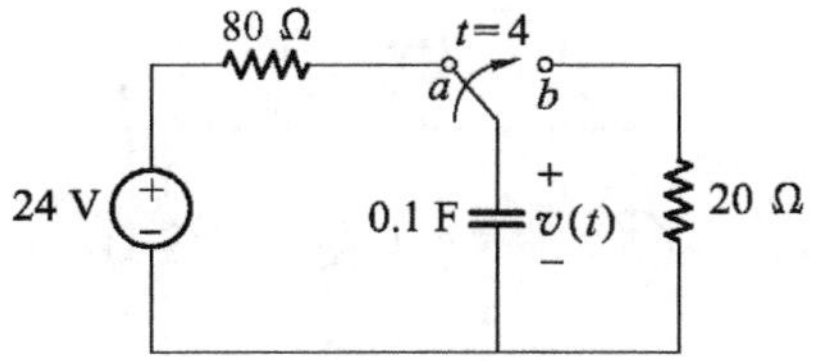

Figure for problem 4. 5

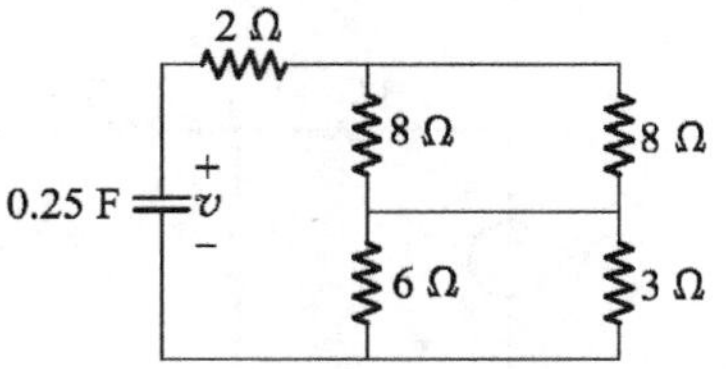

Figure for problem 4. 6

4. 7 Find the time constant for each of the circuits in Figure for problem 4. 7.

4. 8 Consider the circuit of Figure for problem 4. 8. Find $v_0(t)$ if $i(0)=2$ A and $v(t)=0$.

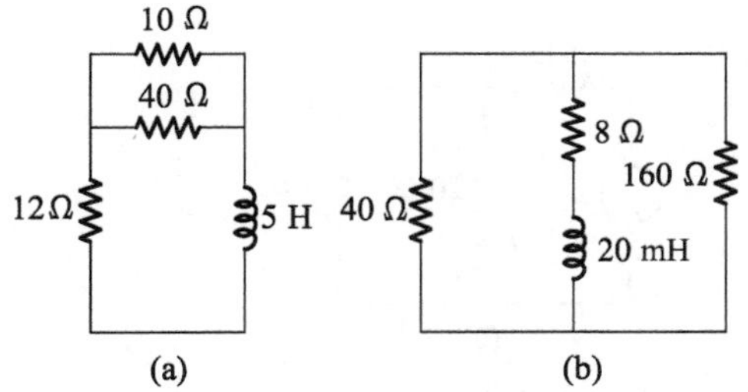

Figure for problem 4. 7

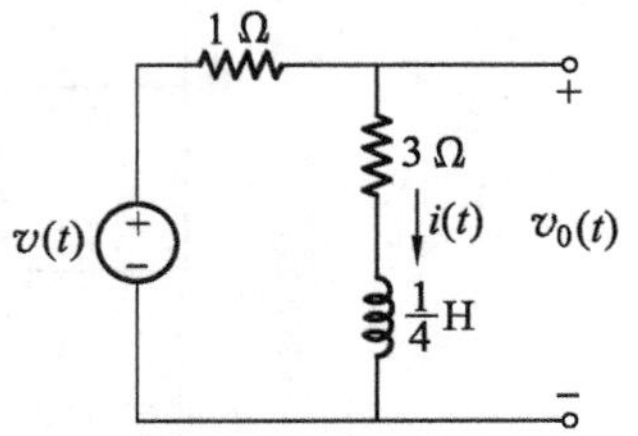

Figure for problem 4. 8

4. 9 Calculate the capacitor voltage for $t<0$ and $t>0$ in each of the circuit in Figure for problem 4. 9.

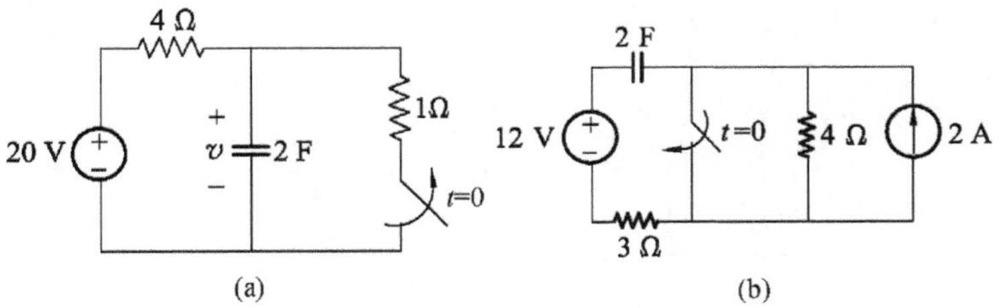

Figure for problem 4. 9

4. 10 In the circuit in Figure for problem 4. 10, find $v(t)$ for $t>0$.

4. 11 Find the step responses $v(t)$ and $i(t)$ to $v_s=5$ V in the circuit of Figure for problem 4. 11.

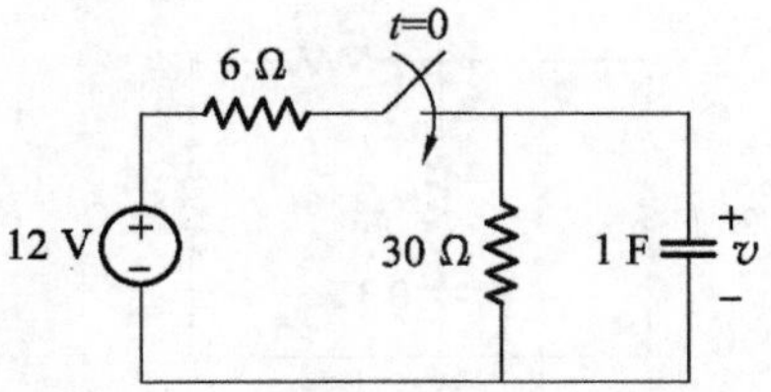

Figure for problem 4.10

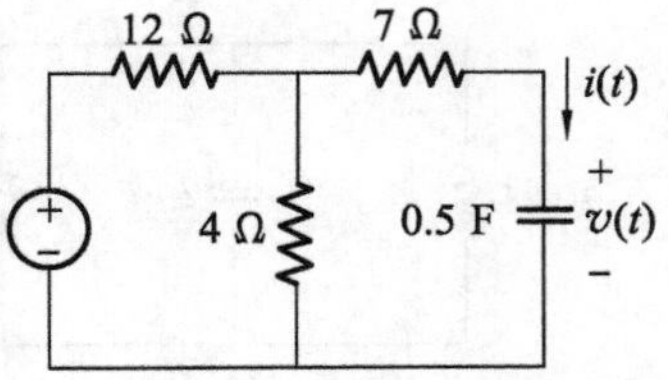

Figure for problem 4.11

4.12 Determine the inductor current $i(t)$ for both $t<0$ and $t>0$ in each of the circuits in Figure for problem 4.12.

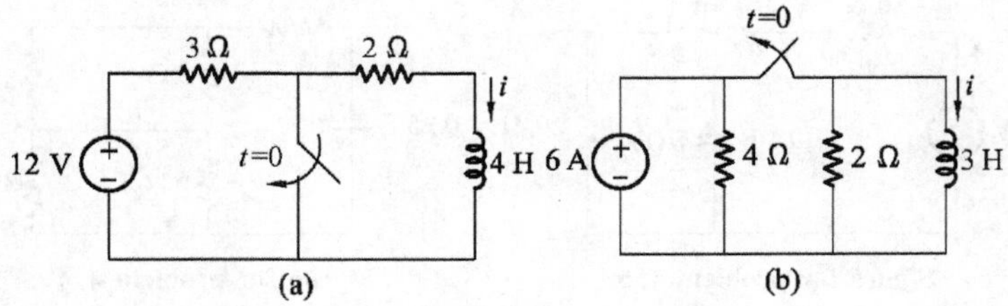

Figure for problem 4.12

4.13 If the input pulse in Figure for problem 4.13(a) is applied to the circuit in Figure for problem 4.13(b), determine the response $i(t)$.

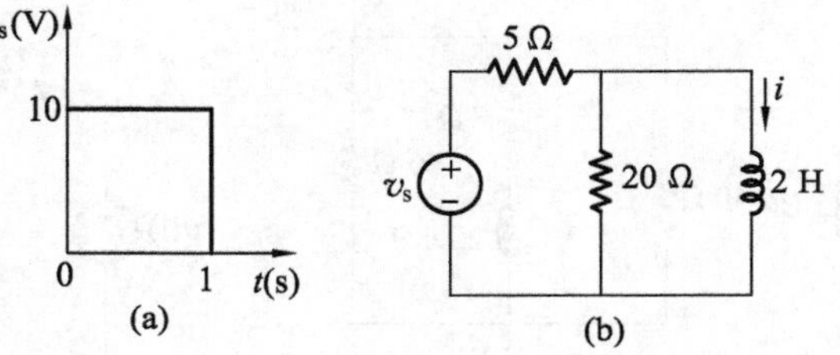

Figure for problem 4.13

CHAPTER 5
SINUSOIDAL STEADY-STATE ANALYSIS

Introduction

Circuits with sinusoidal sources have many important applications. The electric power used for industrial and household applications worldwide is generated and delivered in the form of either 50 or 60 Hz sinusoidal voltages and currents. Furthermore, sinusoidal signals have many uses in radio communication. Finally, a branch of mathematics known as Fourier analysis shows that all signals of practical interest are composed of sinusoidal components. Thus, the study of circuits with sinusoidal sources is a central theme in electrical engineering.

5.1 Sinusoidal currents and voltages

5.1.1 Characters of a sinusoidal signal

A sinusoidal voltage is shown in Figure 5.1.1 and is given by

$$v(t)=V_m\cos(\omega t+\theta) \tag{5.1.1}$$

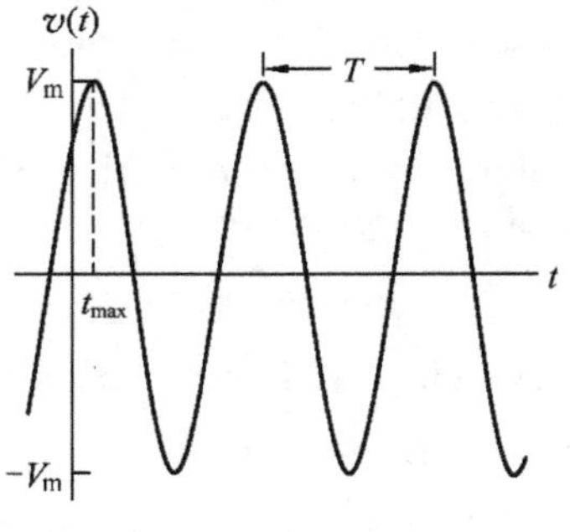

Figure 5.1.1 A sinusoidal voltage waveform

In Figure 5.1.1, V_m is the peak value of the voltage, ω is the angular frequency in radians per second, and θ is the phase angle.

Sinusoidal signals are periodic, repeating the same pattern of values in each period T. Because the cosine (or sine) function completes one cycle when the angle increases by 2π radians, we get

$$\omega T=2\pi \tag{5.1.2}$$

The frequency of a periodic signal is the number of cycles completed in one second. Thus, we obtain

$$f=\frac{1}{T} \tag{5.1.3}$$

The units of frequency are hertz(Hz). (Actually, the physical units of hertz are equivalent to inverse seconds.) Solving Equation (5.1.2) for the angular frequency, we have

$$\omega=\frac{2\pi}{T} \tag{5.1.4}$$

Using Equation (5.1.3) to substitute T, we find that the angular frequency, and that

$$\omega=2\pi f \tag{5.1.5}$$

Throughout our discussion, the argument of the cosine (or sine) function is of the form

$$\omega t+\theta$$

We assume that the angular frequency ω has units of radians per second (rad/s).

However, we sometimes give the phase angle θ in degrees. Then, the argument has mixed units. If we wanted to evaluate $\cos(\omega t+\theta)$ for a particular value of time, we would have to convert ° to radians before adding the terms in the argument. Usually, we find it easier to visualize an angle expressed in degrees, and mixed units are not a problem.

For uniformity, we express sinusoidal functions by using the cosine function rather than the sine function. The functions are related by the identity

$$\sin(z)=\cos(z-90°) \tag{5.1.6}$$

For example, when we want to find the phase angle of

$$v_x(t)=10\sin(200t+30°)$$

we first write it as

$$\begin{aligned} v_x(t)&=10\cos(200t+30°-90°)\\ &=10\cos(200t-60°) \end{aligned}$$

Thus, we state that the phase angle of $v_x(t)$ is $-60°$.

5.1.2 Root-Mean-Square Values

Consider applying a periodic voltage $v(t)$ with period T to a resistance R. The power delivered to the resistance is given by

$$p(t)=\frac{v^2(t)}{R} \tag{5.1.7}$$

Furthermore, the energy delivered in one period is given by

$$E_T=\int_0^T p(t)\,\mathrm{d}t \tag{5.1.8}$$

The average power P_{avg} delivered to the resistance is the energy delivered in one cycle divided by the period. Thus,

$$P_{avg}=\frac{E_T}{T}=\frac{1}{T}\int_0^T p(t)\,\mathrm{d}t \tag{5.1.9}$$

Using Equation (5.1.7) to substitute into Equation (5.1.9), we obtain

$$P_{avg}=\frac{1}{T}\int_0^T \frac{v^2(t)}{R}\mathrm{d}t \tag{5.1.10}$$

Now, we define the root-mean-square (rms) value of the periodic voltage $v(t)$ as

$$V_{rms}=\sqrt{\frac{1}{T}\int_0^T v^2(t)\,\mathrm{d}t} \tag{5.1.11}$$

So, we get

$$P_{avg} = \frac{V_{rms}^2}{R} \tag{5.1.12}$$

Thus, if the rms value of a periodic voltage is known, it is relatively easy to compute the average power that the voltage can deliver to a resistance. The rms value is also called the effective value.

Similarly for a periodic current $i(t)$, we define the rms value as

$$I_{rms} = \sqrt{\frac{1}{T}\int_0^T i^2(t)\mathrm{d}t} \tag{5.1.13}$$

and the average power delivered if $i(t)$ flows through a resistance is given by

$$P_{avg} = I_{rms}^2 R \tag{5.1.14}$$

5.1.3 RMS value of a sinusoid

Consider a sinusoidal voltage given by

$$v(t) = V_m\cos(\omega t + \theta) \tag{5.1.15}$$

To find the rms value, we substitute into Equation(5.1.11), which yields

$$V_{rms} = \sqrt{\frac{1}{T}\int_0^T V_m^2\cos^2(\omega t + \theta)\mathrm{d}t} \tag{5.1.16}$$

Next, we use the trigonometric identity

$$\cos^2(z) = \frac{1}{2} + \frac{1}{2}\cos(2z) \tag{5.1.17}$$

Finally, we get a useful result

$$V_{rms} = \frac{V_m}{\sqrt{2}} \tag{5.1.18}$$

Usually in discussing sinusoids, the rms or effective value is given rather than the peak value. For example, AC power in residential wiring is distributed as a 60 Hz, 115 V rms sinusoid. Here 115 V is the rms value and that the peak value is

$$V_m = \sqrt{2} \times V_{rms} = \sqrt{2} \times 115 \approx 163 \text{ V}$$

Actually, 115 V is the nominal residential distribution voltage. It can vary from approximately 105 to 130 V.

Example 5.1.1 Power delivered to a resistance by a sinusoidal source

Suppose that a voltage given by

$$v(t) = 100\cos(100\pi t) \text{ V}$$

is applied to a 50 Ω resistance. Sketch $v(t)$ to scale versus time. Find the rms value of the voltage and the average power delivered to the resistance. Find the power as a function of time and sketch to scale (Figure for example 5.1.1).

Solution: Here, we see $\omega = 100\pi$, and the frequency is $f = \omega/2\pi = 50$ Hz. Then, the period is $T = 1/f = 20$ ms. A plot of $v(t)$ versus time is shown in Figure for example 5.1.1. The peak value of the voltage is $V_m = 100$ V. Thus, the rms value is

$$V_{rms} = V_m/\sqrt{2} = 70.71\ V$$

Then, the average power is

$$P_{avg} = V_{rms}^2/R = 100\ W$$

The power as a function of time is given by

$$P_{avg} = v^2(t)/R = 100^2\cos^2(100\pi t)/50 = 200\cos^2(100\pi t)\ W$$

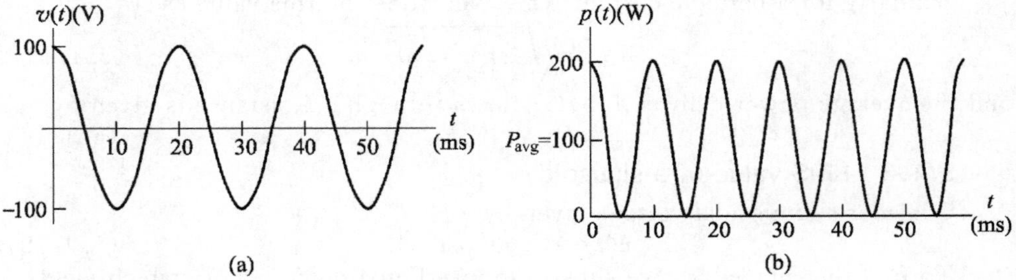

Figure for example 5.1.1

5.2 Phasors

In the next several sections, we'll see that sinusoidal steady-stated analysis is greatly facilitated if the currents and voltages are represented as vectors (called phasors) in the complex-number plane. In preparation for this material, you may need to learn the review of complex-number arithmetic.

5.2.1 Complex number

5.2.1.1 The representation of a complex number

A number that satisfies the equation

$$x^2 + 16 = 0 \tag{5.2.1}$$

is not a real number. Imaginary numbers are introduced to solve equations such as Equation (5.2.1). To deal with imaginary numbers, a new element, j, is added to the number system, having the property

$$j^2 = -1 \text{ or } j = \sqrt{-1} \tag{5.2.2}$$

Using Equation (5.2.2), we can obtain the solutions to Equation (5.2.11) which are $\pm$j4. In mathematics, the symbol i is used for the imaginary unit, but this might be confused with current in electrical engineering. Therefore, the symbol j is used in this book. A complex number can be expressed by different forms as follows.

(1) Algebraic form

A complex number (indicated in boldface notation) is an expression of the form

$$\mathbf{A} = a + jb \tag{5.2.3}$$

where a and b are real numbers. The complex number **A** has a real part, a, and an imaginary part, b, which can be expressed as

$$a = \text{Re}\,\mathbf{A} \qquad b = \text{Im}\,\mathbf{A} \tag{5.2.4}$$

(2) Rectangular form

The complex number *A* can be represented on a rectangular coordinate plane, called the **complex plane**, by interpreting it as a point (a, b). That is, the horizontal coordinate is a in real axis and the vertical coordinate is b in imaginary axis, as shown in Figure 5.2.1.

The complex number $\mathbf{A} = a + \mathrm{j}b$ can also be uniquely located in the complex plane by specifying the distance r along a straight line from the origin and the angle θ, which is between the line and the real axis, as shown in Figure 5.2.1. From the right triangle of Figure 5.2.1, we can see that:

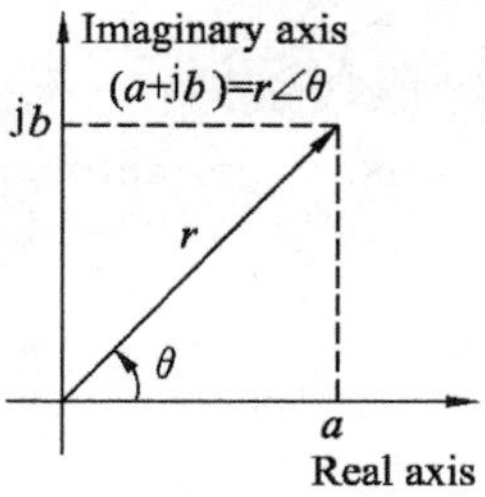

Figure 5.2.1 Polar form representation of complex numbers

$$r = \sqrt{a^2 + b^2}$$
$$\theta = \tan^{-1}\left(\frac{b}{a}\right) \qquad (5.2.4)$$
$$a = r\cos\theta$$
$$b = r\sin\theta$$

Then, we can represent a complex number by the rectangular form:

$$\mathbf{A} = r\cos\theta + \mathrm{j}r\sin\theta \qquad (5.2.5)$$

(3) Exponential form

According to Euler's Identity,

$$\cos\theta = \frac{\mathrm{e}^{\mathrm{j}\theta} + \mathrm{e}^{-\mathrm{j}\theta}}{2},\ \sin\theta = \frac{\mathrm{e}^{\mathrm{j}\theta} - \mathrm{e}^{-\mathrm{j}\theta}}{2\mathrm{j}} \qquad (5.2.6)$$

Thus, a complex number can be represented as the exponential form

$$\mathbf{A} = r\mathrm{e}^{\mathrm{j}\theta} \qquad (5.2.7)$$

(4) Polar form

The polar form of the complex number is

$$\mathbf{A} = r\angle\theta \qquad (5.2.8)$$

In all these forms, the number r is called the magnitude (or amplitude) and the number θ is called the angle (or argument). The two numbers are usually denoted by: $r = |\mathbf{A}|$ and $\theta = \arctan \mathbf{A}$.

In the operation of complex Numbers, these four forms can be converted to each other.

$$\mathbf{A} = a + \mathrm{j}b = r\cos\theta + \mathrm{j}r\sin\theta = r\mathrm{e}^{\mathrm{j}\theta} = r\angle\theta \qquad (5.2.9)$$

Finally, we should remark that two complex numbers are equal only if their magnitudes are equal and their arguments are equal.

Example 5.2.1

Convert the complex number $\mathbf{A} = 6 + \mathrm{j}8$ to its polar form.

Solution:

$$r = \sqrt{6^2 + 8^2} = 10$$
$$\theta = \tan^{-1}\left(-\frac{8}{6}\right) = -53°$$
$$\mathbf{A} = 10\angle -53°$$

Example 5.2.2

Convert the number $\mathbf{A}=4\angle 30^\circ$ to its complex form.

Solution:

$$a=4\cos(30^\circ)=2\sqrt{3}$$
$$b=4\sin(30^\circ)=2$$
$$\mathbf{A}=2\sqrt{3}+j2$$

5.2.1.2 Complex operation

Addition and subtraction of complex numbers take place according to the following rules:

$$(a_1+jb_1)+(a_2+jb_2)=(a_1+a_2)+j(b_1+b_2)$$
$$(a_1+jb_1)-(a_2+jb_2)=(a_1-a_2)+j(b_1-b_2) \quad (5.2.10)$$

Multiplication of complex numbers in polar form follows the law of exponents.

That is, the magnitude of the product is the product of the individual magnitudes, and the angle of the product is the sum of the individual angles, as shown below.

$$(\mathbf{A})(\mathbf{B})=(Ae^{j\theta})(Be^{j\phi})=ABe^{j(\theta+\phi)}=AB\angle(\theta+\phi) \quad (5.2.11)$$

Division of complex numbers in polar form follows the law of exponents. That is, the magnitude of the quotient is the quotient of the magnitudes, and the angle of the quotient is the difference of the angles, as shown in Equation (5.2.12).

$$\frac{\mathbf{A}}{\mathbf{B}}=\frac{Ae^{j\theta}}{Be^{j\phi}}=\frac{A}{B}\angle\theta-\phi \quad (5.2.12)$$

Example 5.2.3

Perform the following operations given that $\mathbf{A}=3+j7$ and $\mathbf{B}=6-j5$. (a) $\mathbf{A}+\mathbf{B}$; (b) $\mathbf{A}-\mathbf{B}$.

Solution:

$$\mathbf{A}+\mathbf{B}=(3+6)+j(7-5)=9+j2$$
$$\mathbf{A}-\mathbf{B}=(3-6)+j[7-(-5)]=-3+j12$$

Example 5.2.4

Perform the following operations given that $\mathbf{A}=3+j4$ and $\mathbf{B}=5+j5$ (a) $\mathbf{AB}$; (b) $\mathbf{A}/\mathbf{B}$.

Solution:

Convert **A** and **B** to their polar forms:

$$\mathbf{A}=5\angle 53^\circ$$
$$\mathbf{B}=5\sqrt{2}\angle 45^\circ$$

Thus

$$\boldsymbol{AB}=25\sqrt{2}\angle 98^\circ$$

$$\boldsymbol{A}/\boldsymbol{B}=\frac{\sqrt{2}}{2}\angle 8^\circ$$

5.2.2 Phasors as rotating vectors

We start with a study of convenient methods for adding (or subtracting) sinusoidal waveforms. We often need to do this in applying Kirchhoff's voltage (KVL) or Kirchhoff's current law (KCL) to AC circuits. For example, in applying KVL to a network with sinusoidal voltages, we might obtain the expression

$$v(t)=10\cos(\omega t)+5\sin(\omega t+60^\circ)+5\cos(\omega t+90^\circ) \qquad (5.2.13)$$

To obtain the peak value of $v(t)$ and its phase angle, we need to change Equation (5.2.13) into the form

$$v(t)=V_m\cos(\omega t+\theta) \qquad (5.2.14)$$

This could be accomplished by repeated substitution, using standard trigonometric identities. However, that method is too tedious for routine work. Instead, we'll use vectors in the complex-number plane known as phasors to represent each term on the right-hand of Equation (5.2.13). Then we can add the phasors with relative ease and convert the sum into the desired form.

Consider a sinusoidal voltage given by

$$v(t)=V_m\cos(\omega t+\theta)$$

In developing the phasor concept, we write

$$v(t)=\mathrm{Re}[V_m e^{j(\omega t+\theta)}] \qquad (5.2.15)$$

The complex quantity inside the brackets is

$$V_m e^{j(\omega t+\theta)}=V_m\angle(\omega t+\theta) \qquad (5.2.16)$$

This can be visualized as a vector of length V_m that rotates counterclockwise in the complex plane with an angular velocity of ω rad/s. Furthermore, the voltage $v(t)$ is the real part of the rector, which is illustrated in Figure 5.2.2. As the vector rotates, its projection on the real axis traces out the voltage as a function of time. The phasor is simply a "snapshot" of this rotating vector at $t=0$.

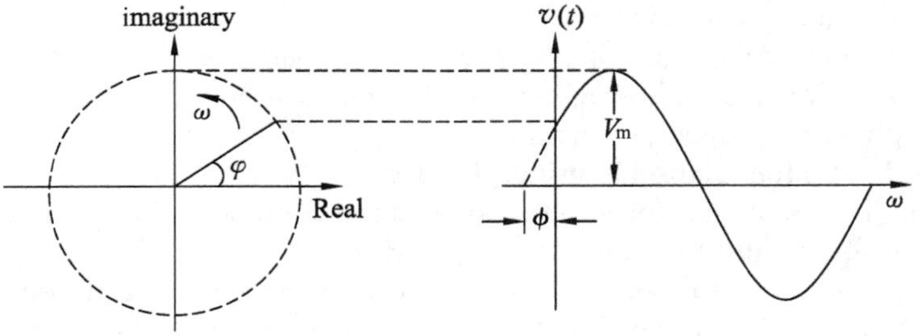

Figure 5.2.2 Phasors as rotating vectors

5.2.3 Phasor definition

Phasors are complex numbers that represent sinusoidal voltages or currents. The magnitude of a phasor equals the peak value and the angle equals the phase of the sinusoid (written as a cosine). We use boldface letters for phasors.

Thus, the **complex phasor** corresponding to the sinusoidal voltage $v_1(t)=V_1\cos(\omega t+\theta_1)$ is defined as

$$\boldsymbol{V}_1=V_{1m}\angle\theta_1 \tag{5.2.17}$$

where V_1 is the peak amplitude of the sinusoid, and θ_1 equals to the phase shift of the sinusoidal signal referenced to a cosine signal.

If the sinusoid is of the form

$$v_2(t)=V_{2m}\sin(\omega t+\theta_2)$$

We first convert it to a cosine function by using the trigonometric identity

Thus, we have

$$v_2(t)=V_{2m}\cos(\omega t+\theta_2-90°)$$

and the phasor is

$$\boldsymbol{V}_2=V_{2m}\angle(\theta_2-90°) \tag{5.2.18}$$

Phasors are obtained for sinusoidal currents in a similar fashion. Thus, for the currents

$$i_1(t)=I_{1m}\cos(\omega t+\theta_1)$$

and

$$i_2(t)=I_{2m}\sin(\omega t+\theta_2)$$

the phasors are

$$\boldsymbol{I}_1=I_{1m}\angle\theta_1 \text{ and } \boldsymbol{I}_2=I_{2m}\angle(\theta_2-90°) \tag{5.2.19}$$

Respectively.

It is important to explicitly point out that this is a definition. Phasor notation arises from Equation (5.2.15); however, this expression is simplified by removing the "real part of" operator (Re) and factoring out and deleting the term $e^{j\omega t}$. The next equation illustrates the simplification:

$$A\cos(\omega t+\varphi)=\mathrm{Re}[Ae^{j(\omega t+\varphi)}]=\mathrm{Re}[Ae^{j\varphi}e^{j\omega t}] \tag{5.2.20}$$

The reason for this simplification is simply for mathematical convenience. You will have to remember that the $e^{j\omega t}$ term that was removed from the complex form of the sinusoid is really still present, indicating the specific frequency of the sinusoidal signal ω.

From now on, we know that any sinusoidal signal may be mathematically represented in one of two ways:

a time-domain form, $v(t)=A\cos(\omega t+\theta)$

and a frequency-domain (or phasor) form, $\boldsymbol{V}(j\omega)=Ae^{j\theta}=A\angle\theta$.

In the remainder of this chapter, bold uppercase quantities will be employed to indicate phasor voltages or currents.

5.2.3.1 Adding sinusoids using phasors

In this discussion, we proceed in small logical steps to illustrate clearly why sinusoids can be adding by adding their phasors.

The first step is to write all the sinusoids as cosine functions by using Equation (5.1.6). Thus, the Equation 5.2.13 can be written as

$$\begin{aligned} v(t)&=10\cos(\omega t)+5\cos(\omega t+60°-90°)+5\cos(\omega t+90°) \\ &=10\cos(\omega t)+5\cos(\omega t-30°)+5\cos(\omega t+90°) \end{aligned} \tag{5.2.21}$$

Referring to Euler's formula, we can write

$$\cos(\theta)=\mathrm{Re}(e^{j\theta})=\mathrm{Re}[\cos(\theta)+j\sin(\theta)] \tag{5.2.22}$$

where the notation Re( ) means that we retain only the real part of the quantity inside the parentheses. Thus, we can rewrite Equation (5.2.21) as

$$v(t)=10\mathrm{Re}[e^{j\omega t}]+5\mathrm{Re}[e^{j(\omega t-30^\circ)}]+5\mathrm{Re}[e^{j(\omega t+90^\circ)}] \quad (5.2.23)$$

When we multiply a complex number $\boldsymbol{Z}$ by a real number A, both the real and imaginary parts of $\boldsymbol{Z}$ are multiplied by A. Thus, Equation (5.2.23) becomes

$$v(t)=\mathrm{Re}[10e^{j\omega t}]+\mathrm{Re}[5e^{j(\omega t-30^\circ)}]+\mathrm{Re}[5e^{j(\omega t+90^\circ)}] \quad (5.2.24)$$

Next, we can write

$$v(t)=\mathrm{Re}[10e^{j\omega t}+5e^{j(\omega t-30^\circ)}+5e^{j(\omega t+90^\circ)}] \quad (5.2.25)$$

because the real part of the sum of several complex quantities is equal to the sum of the real parts. If we factor out the common term $e^{j\omega t}$, Equation (5.2.25) becomes

$$v(t)=\mathrm{Re}[(10+5e^{j-30^\circ}+5e^{j90^\circ})e^{j\omega t}] \quad (5.2.26)$$

Put the complex numbers into polar form, we have

$$v(t)=\mathrm{Re}[(10\angle 0^\circ+5\angle -30^\circ+5\angle 90^\circ)e^{j\omega t}] \quad (5.2.27)$$

Now, we can combine the complex numbers as

$$\begin{aligned}&10\angle 0^\circ+5\angle -30^\circ+5\angle 90^\circ\\&=10+4.33-j2.50+j5\\&=14.5\angle 9.90^\circ\end{aligned} \quad (5.2.28)$$

Using this result in Equation (5.2.27), we have

$$v(t)=\mathrm{Re}[(14.54e^{j9.90^\circ})e^{j\omega t}] \quad (5.2.29)$$

which can be written as

$$v(t)=\mathrm{Re}[14.54e^{j(\omega t+9.90^\circ)}] \quad (5.2.30)$$

Now, using Equation (5.2.22), We can write it as

$$v(t)=14.54\cos(\omega t+9.90^\circ) \quad (5.2.31)$$

Thus, we have changed the original expression for $v(t)$ into the desired form. Notice that the essential part of the work needed to combine the sinusoids is to add the phasors.

5.2.3.2 Procedure for adding sinusoids

From now on, to add sinusoids, we can first write the phasor for each term in the sum, add the phasors by using complex-number arithmetic, and then write the simplified expression for the sum.

Example 5.2.5

Using phasors to calculate

Suppose that

$$v_1(t)=20\cos(\omega t-45^\circ)$$
$$v_2(t)=10\sin(\omega t+60^\circ)$$

Reduce the sum $v_s(t)=v_1(t)+v_2(t)$ to a single term.

Solution: The phasors of $v_1(t)$ and $v_2(t)$ are

$$\mathbf{V}_1(t)=20\angle -45^\circ$$
$$\mathbf{V}_2(t)=10\angle(60^\circ-90^\circ)=10\angle -30^\circ$$

Then, we use complex-number arithmetic to add the phasors and convert the sum to polar form:

$$\begin{aligned}\mathbf{V}_s &= \mathbf{V}_1 + \mathbf{V}_2\\ &= 20\angle -45° + 10\angle -30°\\ &= 14.14 - j14.14 + 8.660 - j5\\ &= 29.77\angle -40.01°\end{aligned}$$

Now, we write the time function corresponding to the phasor $\mathbf{V}_s$.

$$v_s(t) = 29.77\cos(\omega t - 40.01°)$$

5.2.3.3 Phase relationships

We will see that the phase relationships between currents and voltages are often important. Consider the voltages.

$$v_1(t) = 3\cos(\omega t + 40°)$$
$$v_2(t) = 5\cos(\omega t - 60°)$$

The corresponding phasors are

$$\mathbf{V}_1 = 3\angle 40°$$
$$\mathbf{V}_2 = 5\angle -60°$$

The phasor diagram is shown in Figure 5.2.3. Notice that the angle between $\mathbf{V}_1$ and $\mathbf{V}_2$ is 60°: Because the complex vectors rotate counterclockwise, we say that $\mathbf{V}_1$ leads $\mathbf{V}$ by 60°. (An alternative way to state the phase relationship is $\mathbf{V}_2$ lags $\mathbf{V}_1$ by 60°).

The plots of $v_1(t)$ and $v_2(t)$ versus ωt are shown in Figure 5.2.4. Notice that $v_1(t)$ reaches its peak 60° earlier than $v_2(t)$. This is the meaning of the statement that $v_1(t)$ leads $v_2(t)$ by 60°.

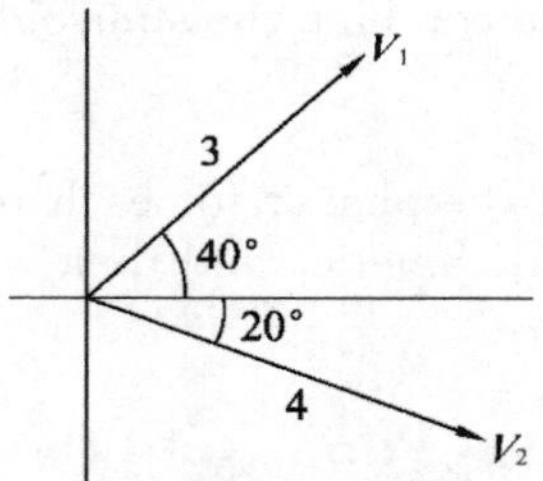

Figure 5.2.3 V_1 leads V_2 by 60°

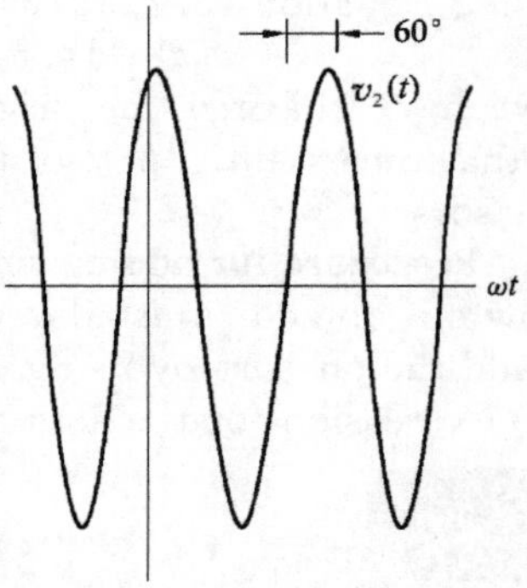

Figure 5.2.4 $v_1(t)$ reaches its peak 60° earlier than $v_2(t)$

5.3 Complex impedances

5.3.1 Inductance

Consider an inductance in which the current is a sinusoid given by

$$i_L = I_m \sin(\omega t + \theta) \tag{5.3.1}$$

The voltage across an inductance is

$$v_L(t) = L\frac{di_L(t)}{dt} \tag{5.3.2}$$

Substituting Equation (5.3.1) into Equation (5.3.2) and reducing, we get

$$v_L(t) = \omega L I_m \cos(\omega t + \theta) \tag{5.3.3}$$

Now the phasors for the current and voltage are

$$\mathbf{I}_L = I_m\angle\theta - 90° \tag{5.3.4}$$

and

$$\mathbf{V}_L = \omega L I_m \angle\theta = V_m \angle\theta \quad (5.3.5)$$

From Equations (5.3.4) and (5.3.5), we get that **the current lags the voltage by 90° for a pure inductance.** The phasor diagram of the current and the voltage is shown in Figure 5.3.1(a). The corresponding waveforms of current and voltage are shown in Figure 5.3.1(b).

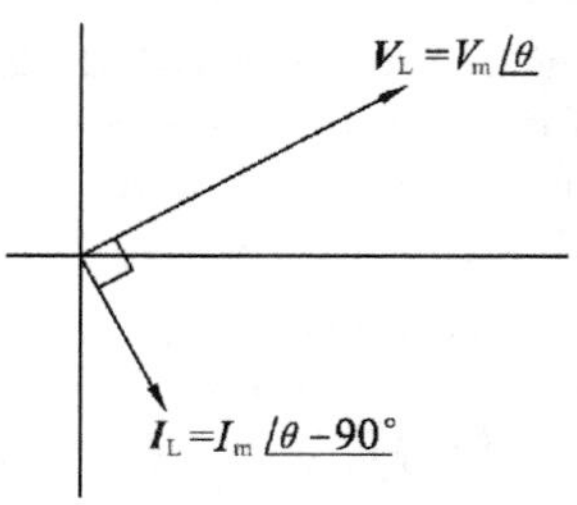

(a) Phasor diagram

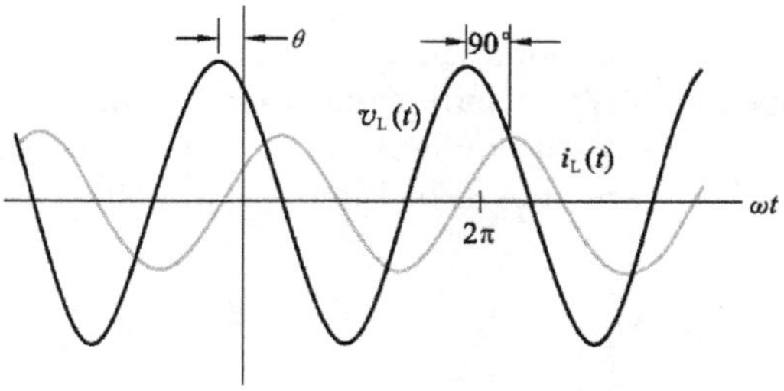

(b) Current and voltage versus time

Figure 5.3.1 Current lays voltage by 90° in a pure inductance

Equation (5.3.5) can be written in the form:

$$\begin{aligned}\mathbf{V}_L &= (\omega L \angle 90°) \times I_m \angle\theta - 90° \\ &= (\omega L \angle 90°) \times \mathbf{I}_L \\ &= j\omega L \times \mathbf{I}_L\end{aligned} \quad (5.3.6)$$

Here, we refer to the term $j\omega L$ as the impedance of the inductance and denote it as X_L. Then we have

$$Z_L = j\omega L = \omega L \angle 90° \quad (5.3.7)$$

and

$$\mathbf{V}_L = Z_L \mathbf{I}_L \quad (5.3.8)$$

From the Equation (5.3.8), we can see that the phasor voltage is equal to the impedance times the phasor current. This is Ohm's law in phasor form. However, for an inductance, the impedance is an imaginary number, whereas resistance is a real number (Impedances that are pure imaginary are also called **reactances.**)

Note that the inductor now appears to behave like a **complex frequency-dependent resistor**, and that the magnitude of this complex resistor, ωL, is proportional to the signal frequency, ω. Thus, an inductor will "impede" current flow in proportion to the sinusoidal frequency of the source signal. This means that at low signal frequencies, an inductor acts somewhat like a short circuit, while at high frequencies it tends to behave more as an open circuit.

5.3.2 Capacitance

Suppose an capacitance in which the voltage is a sinusoid is given by

$$u_C = U_m \sin(\omega t + \theta) \quad (5.3.9)$$

The current across an inductance is

$$i_C(t) = C\frac{du_C(t)}{dt} \quad (5.3.10)$$

Substituting Equation (5.3.9) into Equation (5.3.10) and reducing, we get

$$i_C(t) = \omega C U_m \cos(\omega t + \theta) \quad (5.3.11)$$

Now the phasors for the current and voltage are

$$\boldsymbol{I}_C = \omega C U_m \angle\theta = I_m \angle\theta \tag{5.3.12}$$

and

$$\boldsymbol{U}_C = U_m \angle\theta - 90^\circ \tag{5.3.13}$$

then

$$\boldsymbol{V}_C = -j\frac{1}{\omega C}\boldsymbol{I}_C \tag{5.3.14}$$

From Equations (5.3.13) and (5.3.14), we get that **the current leads the voltage by 90° for a pure capacitance.** The phasor diagram of the current and the voltage are shown in Figure 5.3.2(a). The corresponding waveforms of current and voltage are shown in Figure 5.3.2(b).

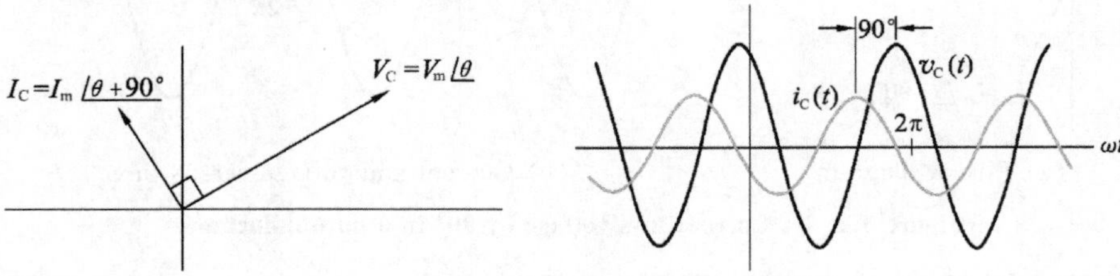

(a) Phasor diagram (b) Current and voltage versus time

Figure 5.3.2 Current leads voltage by 90° in a pure capacitance

Equation (5.3.14) is Ohm's law in phasor form of a capacitance, and the impedance of a capacitance is

$$Z_C = -j\frac{1}{\omega C} = \frac{1}{\omega C}\angle -90^\circ \tag{5.3.15}$$

Now we can show that if the current and voltage of a capacitance are sinusoidal, the phasors are related by

$$\boldsymbol{V}_C = Z_C \boldsymbol{I}_C \tag{5.3.16}$$

Thus, the impedance of a capacitor is also a frequency-dependent complex quantity, with the impedance of the capacitor varying as an inverse function of frequency; and so a capacitor acts like a short circuit at high frequencies, whereas it behaves more like an open circuit at low frequencies.

5.3.3 Resistance

For a resistance, the phasors are related by

$$\boldsymbol{V}_R = \boldsymbol{I}_R R \tag{5.3.17}$$

Because resistance is a real number, the current and voltage are in phase, as illustrated in Figure 5.3.3.

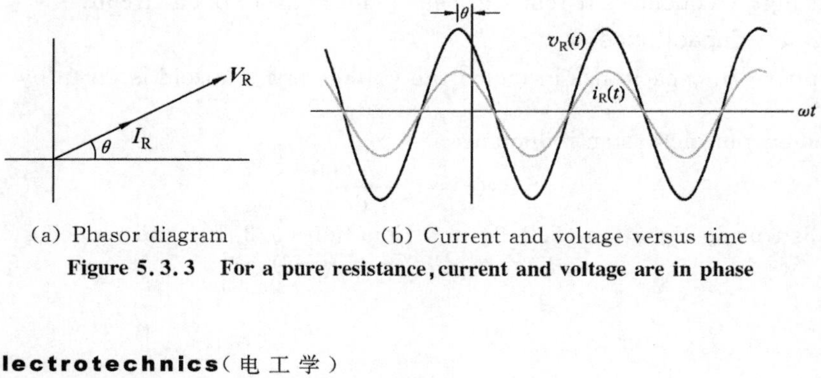

(a) Phasor diagram (b) Current and voltage versus time

Figure 5.3.3 For a pure resistance, current and voltage are in phase

5.3.4 The general form of a complex impedance

Figure 5.3.4 depicts $Z_C(j\omega)$ in the complex plane, alongside $Z_R(j\omega)$ and $Z_L(j\omega)$. The impedance parameter defined in this section is extremely useful in solving AC circuit analysis problems, because it will make it possible to take advantage of most of the network theorems developed for DC circuits by replacing resistances with complex-valued impedances.

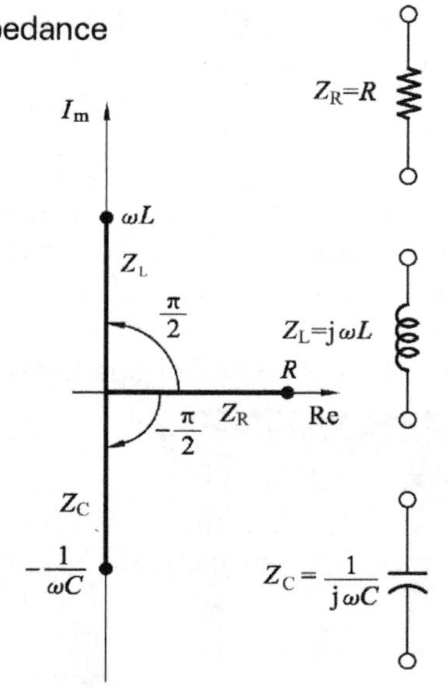

Figure 5.3.4 Impedances of R, L, and C in the complex plane

It is important to emphasize that although the impedance of simple circuit elements is either purely real (for resistors) or purely imaginary (for capacitors and inductors), the general definition of impedance for an arbitrary circuit must allow the possibility of having both a real and an imaginary part, since practical circuits are made up of more or less complex interconnections of different circuit elements. In its most general form, the impedance of a circuit element is defined as the sum of a real part and an imaginary part:

$$Z(j\omega)=R+jX \tag{5.3.18}$$

where R is called the **AC resistance**; and X is called the **reactance**, which is the sum of Z_L and Z_C.

Note that the reactances of Equations (5.3.8) and (5.3.15) have units of Ohms, and that **inductive reactance** is always positive, while **capacitive reactance** is always negative.

5.4 Combinations of complex Impedances

5.4.1 Series combination of complex impedances

Example 5.4.1

Find the equivalent complex impedance in the circuit shown in Figure for example 5.4.1.

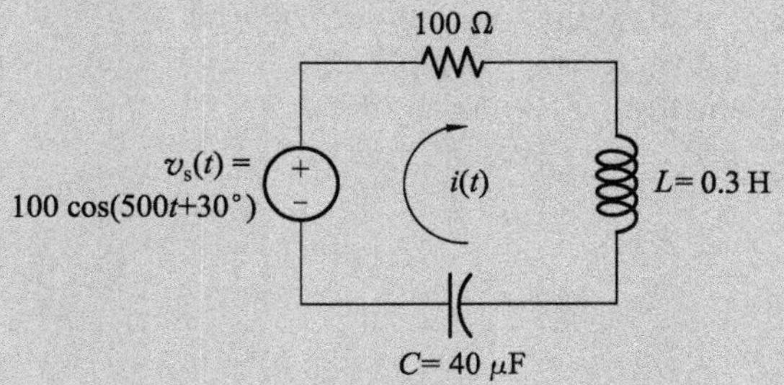

Figure for example 5.4.1

Solution: 【Step 1】 Replace the inductance and the capacitance with their complex impedances.

$$Z_L = j\omega L = j500\times 0.3 = j150\ \Omega$$

$$Z_C = -j\frac{1}{\omega C} = -j\frac{1}{500\times 40\times 10^{-6}} = -j50\ \Omega$$

【Step 2】 Find the equivalent impedance of the circuit by adding the impedances of all three elements. Because all three elements are in series, the equivalent impedance of the circuit is

$$Z_{eq} = R + Z_L + Z_C = 100 + j100 = 141.4\angle 45°\ \Omega$$

5.4.2 Parallel combination of complex impedances

Example 5.4.2

Find the equivalent complex impedance in the circuit shown in Figure for example 5.4.2.

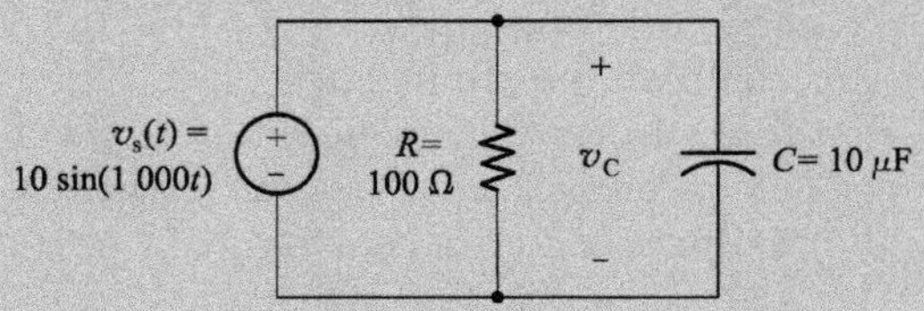

Figure for example 5.4.2

Solution: 【Step1】 Replace the capacitance with its complex impedance.

$$Z_C = -j\frac{1}{\omega C} = -j\frac{1}{1\,000\times 10\times 10^{-6}} = -j100\ \Omega$$

【Step2】 The equivalent impedance of the circuit is

$$Z_{eq} = \frac{1}{\frac{1}{R}+\frac{1}{Z_C}} = \frac{1}{\frac{1}{100}+\frac{1}{-j100}} = 70.71\angle -45°\ \Omega$$

5.4.3 Impedance of a more complex circuit

Example 5.4.3

Find the equivalent impedance of the circuit shown in Figure for example 5.4.3.

Known quantities: $\omega = 104$ rad/s; $R_1 = 100\ \Omega$; $L = 10$ mH; $R_2 = 50\Omega$; $C = 10\ \mu F$.

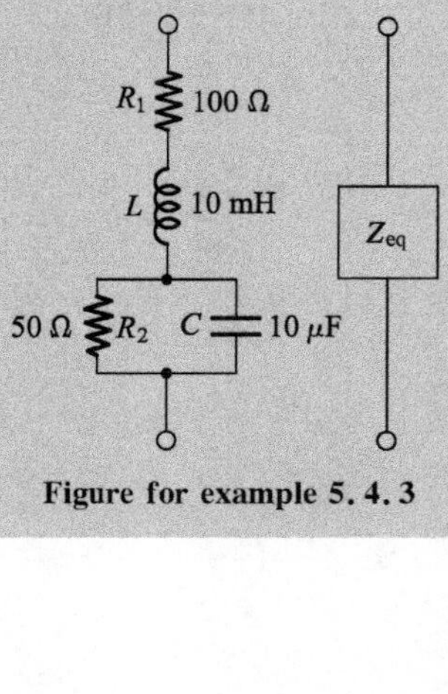

Figure for example 5.4.3

Solution: We determine first the parallel impedance of the R_2, C circuit, $Z_{\parallel}$.

$$Z_{\parallel}=R_2 \parallel \frac{1}{j\omega C}=\frac{R_2\,\dfrac{1}{j\omega C}}{R_2+\dfrac{1}{j\omega C}}=\frac{R_2}{1+j\omega CR_2}$$

$$=\frac{50}{1+j10^4\times 10\times 10^{-6}\times 50}=\frac{50}{1+j5}=1.92-j9.62$$

$$=9.81\angle -78.71^\circ\ \Omega$$

Next, we determine the equivalent impedance, Z_{eq}:

$$Z_{eq}=R_1+j\omega L+Z_{\parallel}=100+j10^4\times 10^{-2}+1.92-j9.62$$

$$=101.92+j90.38=136.2\angle 41.57^\circ\ \Omega$$

Comments: At the frequency used in this example, the circuit has an inductive impedance, since the reactance is positive (or, alternatively, the phase angle is positive).

5.5 Circuits analysis with phasors and complex impedances

5.5.1 Kirchhoff's law in phasor form

Recall that KVL requires that the voltages sum to zero for any closed path in an electrical network. A typical KVL equation is

$$v_1(t)+v_2(t)+v_3(t)+\cdots v_n(t)=0 \tag{5.5.1}$$

If the voltages are sinusoidal, they can be represented by phasors. Then the Equation (5.5.1) becomes

$$\mathbf{V}_1+\mathbf{V}_2+\mathbf{V}_3+\cdots+\mathbf{V}_n=\mathbf{0} \tag{5.5.2}$$

Thus, we can apply KVL directly to the phasors, and the sum of the phasor voltages equals zero for any closed path.

Similarly, KCL can be applied to currents in phasor form. The sum of the phasor currents entering a node must equal the sum of the phasor currents leaving.

5.5.2 A step-by-step procedure for steady-state analysis of circuits with sinusoidal sources

(1) Replace the time descriptions of the voltage and current sources with the corresponding phasors. (All of the sources must have the same frequency.)

(2) Replace inductances by their complex impedances $Z_L= j\omega L$.

(3) Replace capacitances by their complex impedances $Z_C=1/(j\omega C)$.

(4) Resistances have impedances equal to their resistances.

(5) Analyze the circuit by using any of the techniques studied, and perform the calculations with complex arithmetic.

Example 5.5.1

Find the steady-state current and the phasor voltage across each element and construct a phasor diagram for the circuit shown in Figure A for example 5.5.1.

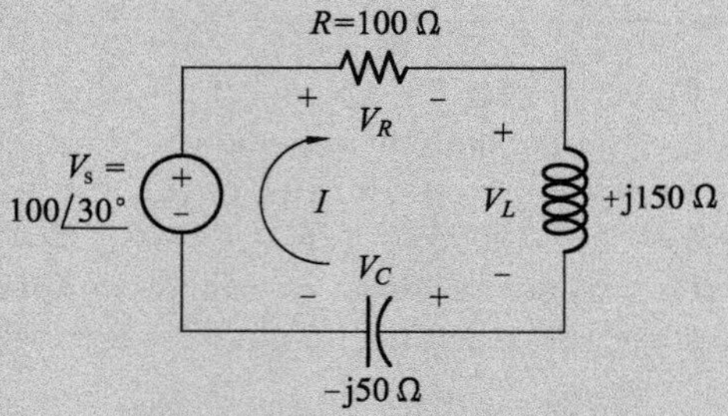

Figure A for example 5.5.1

Solution: The phasor for the voltage source is $\boldsymbol{V}_s = 100\angle 30°$ V, and the transformed circuit is shown in Figure A for example 5.5.1. The equivalent impedance of the circuit is shown in example 5.4.1:

$$Z_{eq} = R + Z_L + Z_C = 100 + j100 = 141.4\angle 45°\ \Omega$$

Now, we can get the phasor current

$$\boldsymbol{I} = \frac{\boldsymbol{V}_s}{Z_{eq}} = \frac{100\angle 30°}{141.4\angle 45°} = 0.707\angle -15°\ \text{A}$$

As a function of time, the current is

$$i(t) = 0.707\cos(500t - 15°)\ \text{A}$$

Next, we can find the phasor voltage across each element by multiplying the phasor current by the respective impedance:

$$\boldsymbol{V}_R = R \times \boldsymbol{I} = 100 \times 0.707\angle -15° = 70.7\angle -15°\ \text{V}$$

$$\boldsymbol{V}_L = Z_L \times \boldsymbol{I} = 150\angle 90° \times 0.707\angle -15° = 106.1\angle 75°\ \text{V}$$

$$\boldsymbol{V}_R = Z_C \times \boldsymbol{I} = 50\angle -90° \times 0.707\angle -15° = 35.4\angle -105°\ \text{V}$$

The phasor diagram for the current and voltages is shown in Figure B for example 5.5.1. Notice that the current $\boldsymbol{I}$ lags the source voltage $\boldsymbol{V}_s$ by 45°. As we can see, the voltage $\boldsymbol{V}_R$ and current $\boldsymbol{I}$ are in phase for the resistance. For the inductance, the voltage $\boldsymbol{V}_L$ leads the current $\boldsymbol{I}$ by 90°.

For the capacitance, the voltage $\boldsymbol{V}_C$ lags the current by 90°.

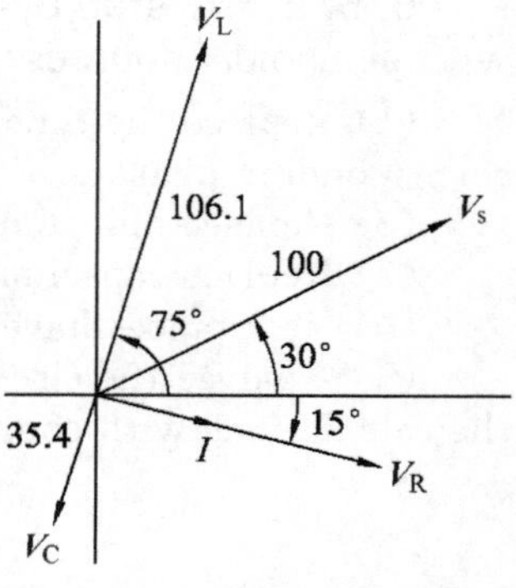

Figure B for example 5.5.1

Example 5.5.2

Consider the circuit shown in Figure A for example 5.5.2(a). Find the voltage $v_C(t)$ in steady state. Find the phasor current through each element, and construct a phasor diagram showing the currents and the source voltage.

Solution: The phasor for the voltage source is $\mathbf{V}_s = 10\angle -90°$ V. The angular frequency of the source is $\omega = 1\ 000$ rad/s. The impedances of the inductance and capacitance are

$$Z_L = j\omega L = j1\ 000 \times 0.1 = j100\ \Omega$$

$$Z_C = -j\frac{1}{\omega C} = -j\frac{1}{1\ 000 \times 10 \times 10^{-6}} = -j100\ \Omega$$

The transformed network is shown in Figure A for example 5.5.2(b).

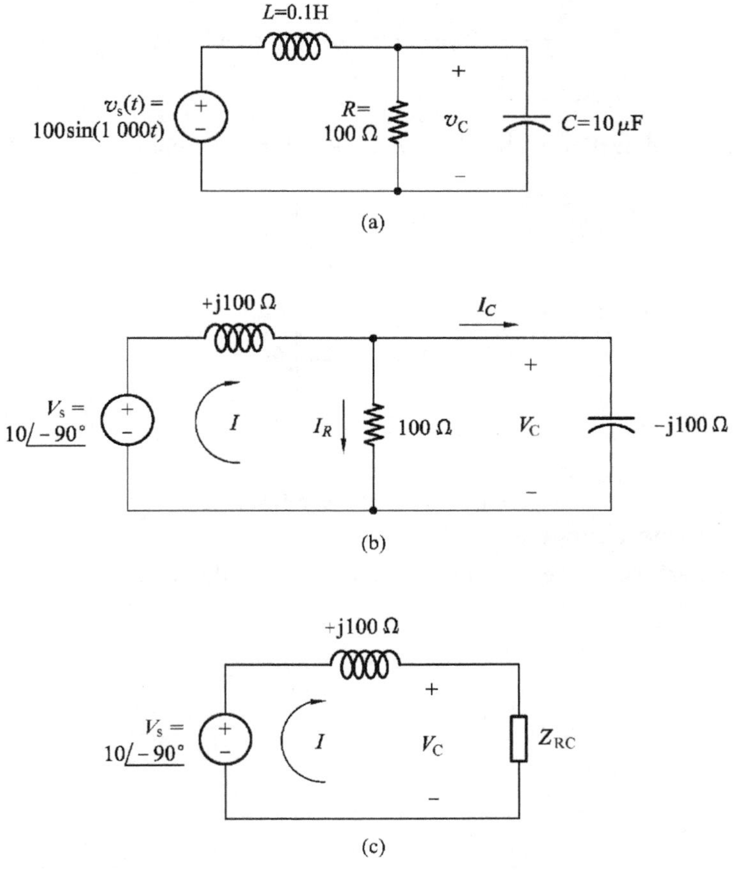

Figure A for example 5.5.2

To find $\mathbf{V}_C$, we will first combine the resistance and the impedance of the capacitor in parallel. Then, we will use the voltage-division principle to compute the voltage across the RC combination. The impedance of the parallel RC circuit is computed in Example 5. 4. 2

$$Z_{RC}=\frac{1}{\frac{1}{R}+\frac{1}{Z_C}}=\frac{1}{\frac{1}{100}+\frac{1}{-j100}}=70.71\angle-45^\circ=50-j50\ \Omega$$

The equivalent network is shown in Figure A for example 5. 5. 2(c).

Now, we can use the voltage-division principle to get

$$\mathbf{V}_C=\mathbf{V}_S\ \frac{Z_{RC}}{Z_L+Z_{RC}}=10\angle-90^\circ\frac{70.71\angle-45^\circ}{j100+50-j50}=10\angle-180^\circ\ \text{V}$$

Converting the phasor to a time function, we have

$$v_C(t)=10\cos(1\,000t-180^\circ)=-10\cos(1\,000t)\ \text{V}$$

Next, we compute the current in each element yielding

$$\boldsymbol{I}=\frac{\mathbf{V}_s}{Z_L+Z_{RC}}=\frac{10\angle-90^\circ}{j100+50-j50}=0.14\angle-135^\circ\ \text{A}$$

$$\boldsymbol{I}_R=\frac{\mathbf{V}_C}{R}=\frac{10\angle-180^\circ}{100}=0.1\angle-180^\circ\ \text{A}$$

$$\boldsymbol{I}_C=\frac{\mathbf{V}_C}{Z_C}=\frac{10\angle-180^\circ}{-j100}=0.1\angle-90^\circ\ \text{A}$$

The phasor diagram is shown in Figure B for example 5. 5. 2.

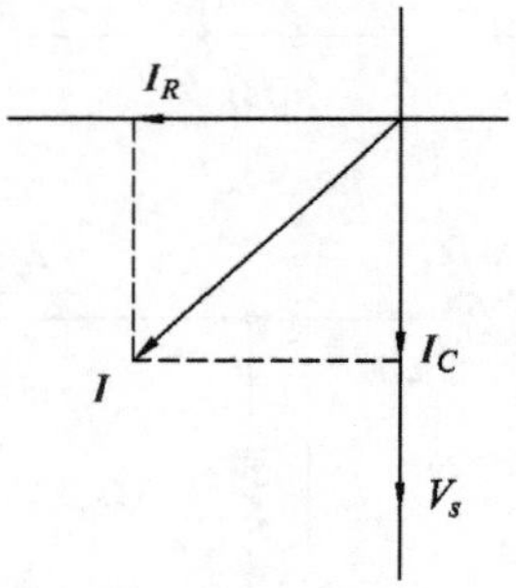

Figure B for example 5. 5. 2

5.5.3 Node-voltage analysis

We can perform node-voltage to analyze a steady-state AC circuit by using phasors.

Example 5.5.3

Use the node-voltage technique to find $v_1(t)$ in steady state in the circuit shown in Figure for example 5.5.3(a).

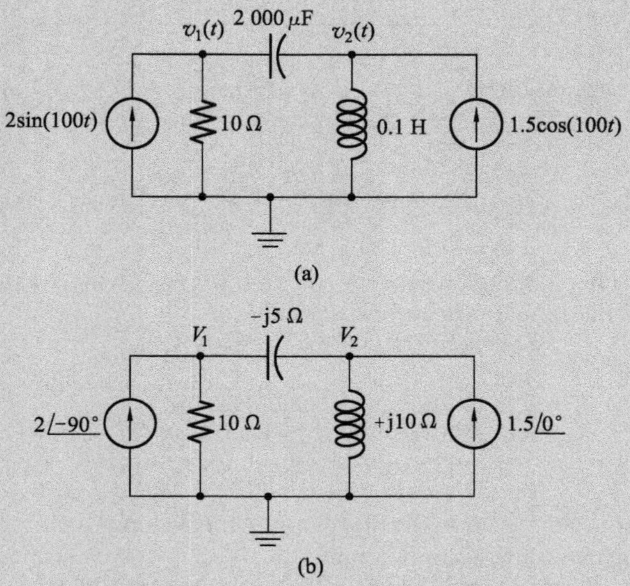

Figure for example 5.5.3

Solution: The transformed network is shown in Figure for example 5.5.3 (b). Using KCL at node 1 and node 2 we can get two equations

$$\frac{\mathbf{V}_1}{10}+\frac{\mathbf{V}_1-\mathbf{V}_2}{-\mathrm{j}5}=2\angle-90°$$

$$\frac{\mathbf{V}_2}{\mathrm{j}10}+\frac{\mathbf{V}_2-\mathbf{V}_1}{-\mathrm{j}5}=1.5\angle0°$$

Change them into the standard form

$$(0.1+\mathrm{j}0.2)\mathbf{V}_1-\mathrm{j}0.2\mathbf{V}_2=-\mathrm{j}2$$

$$-\mathrm{j}0.2\mathbf{V}_1+\mathrm{j}0.1\mathbf{V}_2=1.5$$

Now, we solve for $\mathbf{V}_1$ yielding

$$\mathbf{V}_1=16.1\angle29.7°$$

Then, we can change the phasor to a time function and get

$$v_1(t)=16.1\cos(100t+29.7°)$$

5.5.4 Mesh-current analysis

We can also use phasors to carry out mesh-current analysis in AC circuits.

Example 5.5.4

Find the mesh currents shown in Figure for example 5.5.4.

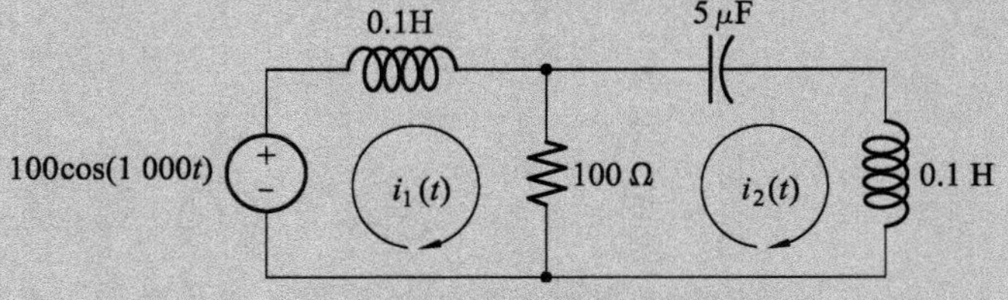

Figure for example 5.5.4

Solution: The angular frequency of the source is $\omega = 1\ 000$ rad/s. The impedances of the inductance and capacitance are

$$Z_L = j\omega L = j1\ 000 \times 0.1 = j100\ \Omega$$

$$Z_C = -j\frac{1}{\omega C} = -j\frac{1}{1\ 000 \times 5 \times 10^{-6}} = -j200\ \Omega$$

$$j100\boldsymbol{I}_1 + 100(\boldsymbol{I}_1 - \boldsymbol{I}_2) = 100\angle 0°$$

$$-j200\boldsymbol{I}_2 + j100\boldsymbol{I}_2 - 100(\boldsymbol{I}_1 - \boldsymbol{I}_2) = 0$$

We get

$$\boldsymbol{I}_1 = 1.414\angle -45°\ \text{A};\ \boldsymbol{I}_2 = 1\ \text{A}$$

Put the phasors into time function

$$i_1 = 1.414\cos(1\ 000t - 45°)\ \text{A};\ i_2 = \cos(1\ 000t)\ \text{A}$$

5.5.5 Thevenin and Norton equivalent circuit

In previous study we learned that a two-terminal network composed of sources and resistances has a Thevenin equivalent circuit consisting of a voltage source in series with a resistance. We can apply this concept to circuits composed of sinusoidal sources (all having a common frequency), resistances, inductances, and capacitances.

5.5.5.1 Thevenin equivalent circuit

Here, the Thevenin equivalent consists of a phasor voltage source in series with a complex impedance as shown in Figure for example 5.5.1. Recall that phasors and complex impedances apply only for steady-state operation; therefore, these Thevenin equivalents are valid for only steady-state operation of the circuit.

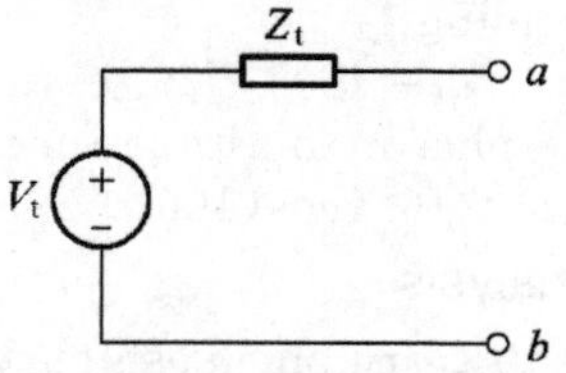

Figure 5.5.1 The Thevenin equivalent for an AC circuit consists of a phasor voltage source V_t in series with a complex impedance Z_t

As in resistive circuits, the Thevenin voltage is equal to the open-circuit voltage of the two-terminal circuit. In AC circuits, we use phasors, so we can write

$$\boldsymbol{V}_{t}=\boldsymbol{V}_{OC}$$

The Thevenin impedance Z_t can be found by zeroing all the independent sources and looking back into the terminals to find the equivalent impedance. (Recall that in zeroing a voltage source, we reduce its voltage to zero, and it becomes a short circuit. On the other hand, in zeroing a current source, we reduce its current to zero, and it becomes an open circuit.) Also, keep in mind that we must not zero the dependent sources.

Another approach to determining the Thevenin impedance is first to find the short-circuit phasor current $\boldsymbol{I}_{SC}$ and the open-circrut voltage $\boldsymbol{V}_{OC}$. Then, the The venin impedance is given by

$$\boldsymbol{Z}_{t}=\frac{\boldsymbol{V}_{OC}}{\boldsymbol{I}_{SC}}$$

Thus, except for the use of phasors and complex impedances, the concepts and procedure for Thevenin equivalents of steady-state AC circuits are the same as for resistive circuits.

5.5.5.2 Norton equivalent circuit

Another equivalent for a two terminal steady-state ac circuit is the Norton equivalent, which consists of a phasor current source $\boldsymbol{I}_n$ in parallel with the The venin impedance. This is shown in Figure 5.5.2. The Norton current is equal to the short-circuit current of the original circuit:

$$\boldsymbol{I}_{n}=\boldsymbol{I}_{SC}$$

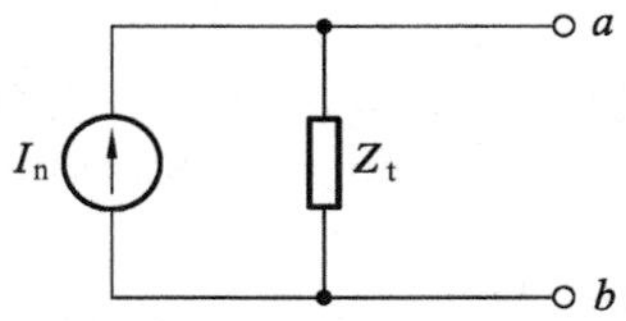

Figure 5.5.2 The Norton equivalent circuit consists of a phasor current source $\boldsymbol{I}_n$ in paralled with the complex impedance Z_t

Example 5.5.5

Find the Thevenin and Norton equivalent circuits for the circuit shown in Figure A for example 5.5.5(a).

Solution: We must find $\boldsymbol{V}_{OC}$, $\boldsymbol{I}_{SC}$, and $\boldsymbol{Z}_t$.

【Step 1】 Zero the sources to find $\boldsymbol{Z}_t$.

If we zero the sources, we obtain the circuit shown in Figure A for example 5.5.5(b). The Thevenin impedance is the impedance looked back into terminals *a-b*. Obviously, the resistance and the capacitance are in parallel. Thus, we get

$$Z_t=\frac{1}{\frac{1}{100}+\frac{1}{-\text{j}100}}=\frac{1}{0.01+\text{j}0.01}=50-\text{j}50\ \Omega$$

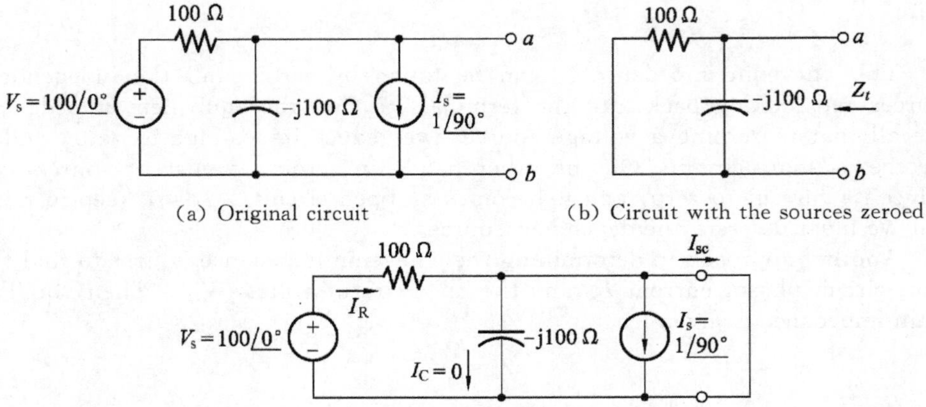

(a) Original circuit (b) Circuit with the sources zeroed

(c) Circuit with a short circuit

Figure A for example 5. 5. 5

Now, we apply a short circuit to terminals a-b and find the current, which is shown in Figure A for example 5. 5. 5(c). With a short-circuit, the voltage across the capacitance is zero. Therefor, $\boldsymbol{I}_C = 0$.

Furthermore, $\boldsymbol{I}_R = \dfrac{\boldsymbol{V}_s}{100} = 1\angle 0°$ A.

Then applying KCL, we can get $\boldsymbol{I}_{SC} = \boldsymbol{I}_R - \boldsymbol{I}_S = 1 - 1\angle 90° = 1.414\angle -45°$ A. Next we can get

$$\boldsymbol{V}_t = \boldsymbol{I}_{SC} Z_t = 1.414\angle -45° \times 70.71\angle -45° = 100\angle -90° \text{ V}$$

Finally, we can draw the Thevenin and Norton equivalent circuits shown in Figure B for example 5. 5. 5.

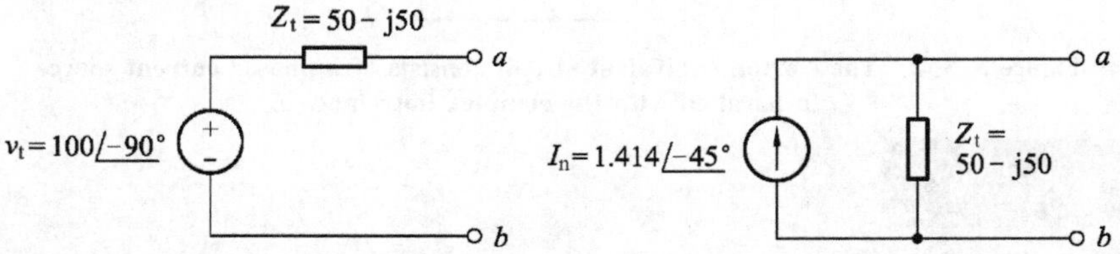

(a) Thevenin equivalent (b) Norton equivalent

Figure B for example 5. 5. 5

5.6 Simulate the sinusoidal response for RLC circuits using Multisim

To RLC series circuits, when we input a sinusoidal signal, if we change the value of capacitor, the circuits may be inductive or capacitive. The variation of the phase angles between current and voltage can be simulated by Multisim.

Find corresponding components and instruments in component library by Multisim and connect them to a circuit as shown in Figure 5. 6. 1. The capacitor C1 in this circuit is a variable capacitance. The increment is set to be 0. 5%. We will measure the total power dissipation and the power factor using Wattmeter of Multisim.

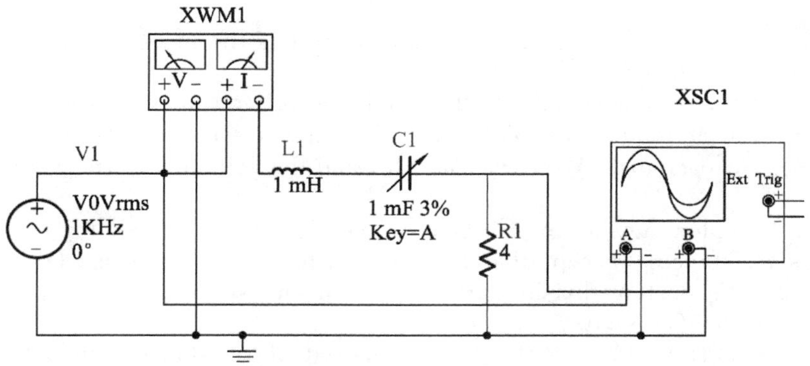

Figure 5. 6. 1 RLC series circuit in Multisim

The simulating waveform and the measured values are shown in Figure 5. 6. 2. When the increment is 2. 5%, the power dissipation is maximum, and the power factor is 0. 999 79. Total voltage and current are practically in phase. Thus, the circuit is almost in resonant state.

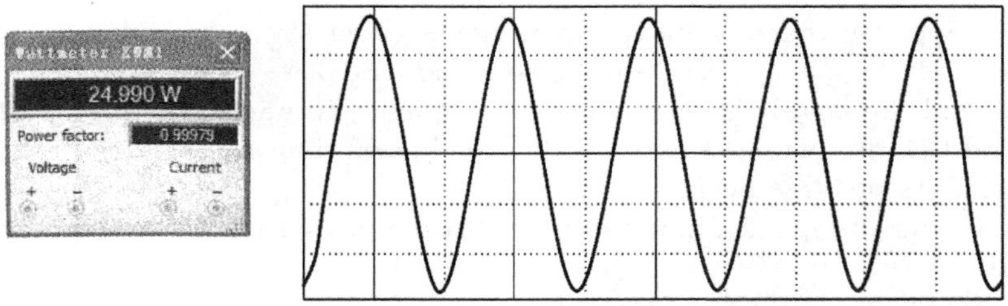

Figure 5. 6. 2 The result of simulation

Summary

1. A sinusoidal voltage is given by $v(t)=V_m \cos(\omega t+\theta)$, where V_m is the peak value of the voltage, ω is the angular frequency in radians per second, and θ is the phase angle. The frequency in hertz is $f=1/T$, where T is the period. Furthermore, $\omega=2\pi f$.

2. For uniformity, we express sinusoidal voltages in terms of the cosine function. A sine function can be converted to a cosine function by use of the identity $\sin(z)=\cos(z-90°)$.

3. For a sinusoid, the root-mean-square (rms) value (or effective value) of a periodic voltage v(t) is the peak value divided by $\sqrt{2}$.

4. We can represent sinusoids with phasors. The magnitude of the phasor is the peak value of the sinusoid. The phase angle of the phasor is the phase angle of the sinusoid (assuming that we have written the sinusoid in terms of a cosine function).

5. The phasor voltage for a passive circuit is the phasor current times the complex impedance of the circuit.

6. For a resistance, $\mathbf{V}_R = R\mathbf{I}_R$, and the voltage is in phase with the current. For an inductance, $\mathbf{V}_L = j\omega L\mathbf{I}_L$, and the voltage leads the current by 90°. For a capacitance, $\mathbf{V}_C = -j(1/\omega C)\mathbf{I}_C$, and the voltage lags the current by 90°.

7. Complex impedances can be combined in series or parallel in the same way as resistances (except that complex arithmetic must be used). Node-voltages, the current-division principle, and the voltage-division principle also apply to AC circuits.

8. In steady state, a network composed of resistances, inductances, capacitances, and sinusoidal sources (all of the same frequency) has a Thevenin equivalent consisting of a phasor voltage source in series with a complex impedance. The Norton equivalent consists of a phasor current source in parallel with the Thevenin impedance.

Problems

5.1 Suppose that a sinusoidal voltage is given by

$$v(t) = 150\cos(200\pi t - 30°)\ \text{V}$$

(a) Find the angular frequency, the frequency in hertz, the period, the peak value, and the rms value. Also, the first value of time t_{max} after $t=0$ such that $v(t)$ attains its positive peak.

(b) If this voltage is applied to a 50 Ω resistance, compute the average power delivered.

(c) Sketch $v(t)$ to scale versus time.

5.2 Express $v(t) = 100\sin(300\pi t + 60°)$ V as a cosine function.

5.3 Consider the circuit shown in Figure for problem 5.3(a).

(a) Find $i(t)$.

(b) Construct a phasor diagram showing all three voltages and the current.

(c) What is the phase relationship between $v_s(t)$ and $i(t)$?

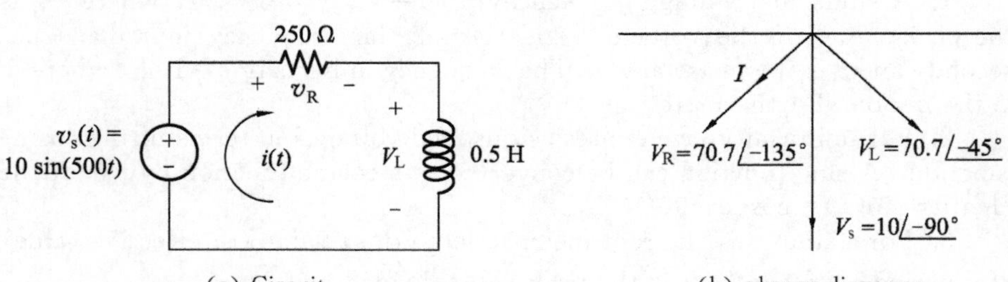

(a) Circuit (b) phasor diageram

Figure for problem 5.3

5.4 Find the phasor voltage and the phasor current through each element in the circuit of Figure for problem 5.4.

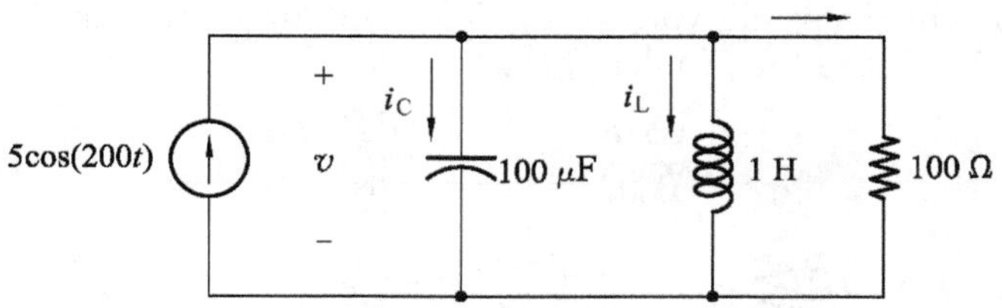

Figure for problem 5. 4

5. 5 Consider the voltages given by

$$v_1(t)=10\cos(\omega t-30°)$$
$$v_2(t)=10\cos(\omega t+30°)$$
$$v_3(t)=10\sin(\omega t+45°)$$

State the phase relationship between each pair of the voltages.

5. 6 Find $v_{out}(t)$ for the circuit shown in Figure for problem 5. 6

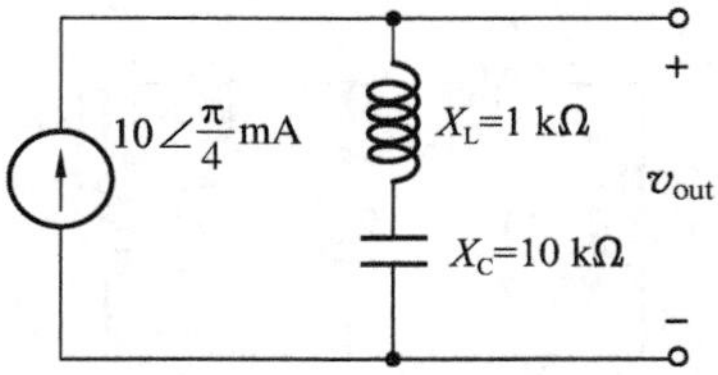

Figure for problem 5. 6

5. 7 Find the phasor values of $\mathbf{V}_R$, $\mathbf{V}_L$ and $\mathbf{V}_C$ in polar form in the circuit of Figure for problem 5. 7.

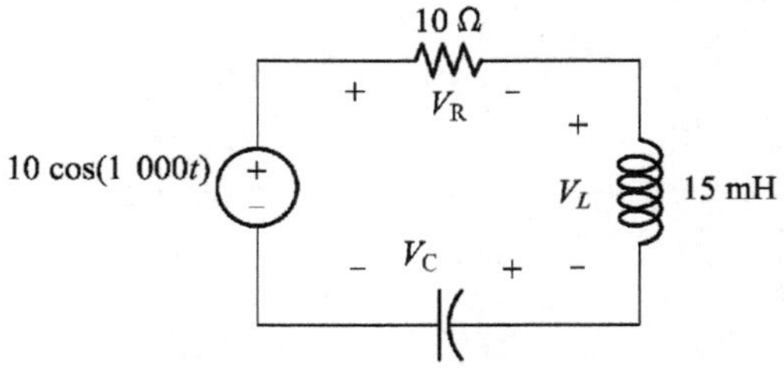

Figure for problem 5. 7

5. 8 Solve for the mesh currents shown in Figure for problem 5. 8.

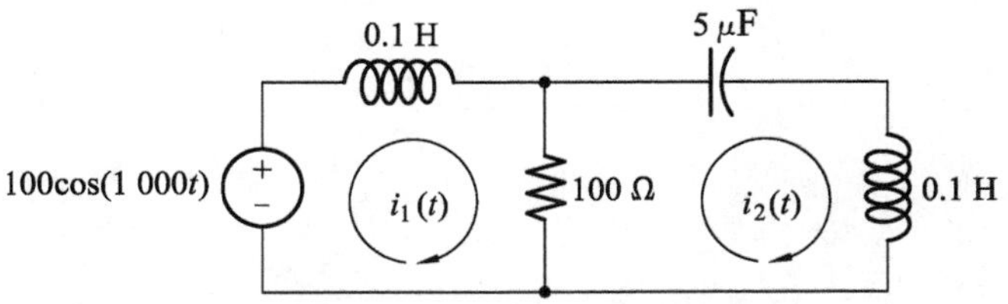

Figure for problem 5. 8 Circuit for problem 5. 8

5.9 Find the Thevenin voltage and Thevenin impedance in the two-terminal circuit shown in Figure for problem 5.9.

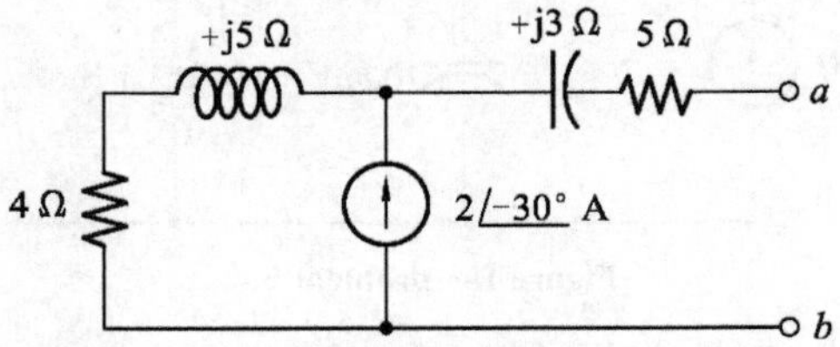

Figure for problem 5.9

5.10 Compute the Thevenin voltage seen by resistor R_2 in Figure for problem 5.10.

If $v_s(t)=2\cos(2t)$, $R_1=4\ \Omega$, $R_2=4\ \Omega$, $L=2$ H, and $C=\frac{1}{4}$F.

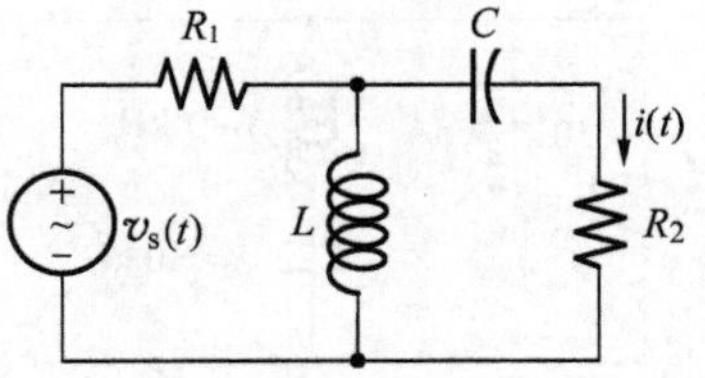

Figure for problem 5.10

CHAPTER 6
AC POWER

Introduction

The objective of this chapter is to introduce the notion of AC power. Before we consider the power delivered by the source to a general load, it is instructive to consider a pure resistive load, a pure inductive load, and a pure capacitive load. The phasor notation developed in Chapter 5 will be employed to analyze the power absorbed by both resistive and complex loads.

6.1 Instantaneous power

From last chapter, you already know that when a linear electric circuit is excited by a sinusoidal source, all voltages and currents in the circuit are also sinusoids of the same frequency as that of the excitation source. The most general expressions for the voltage and current delivered to an arbitrary load are as follows

$$\begin{aligned} v(t) &= V_m \cos(\omega t + \theta_V) \\ i(t) &= I_m \cos(\omega t + \theta_I) \end{aligned} \tag{6.1.1}$$

where V_m and I_m are the peak amplitudes of the sinusoidal voltage and current, respectively, and θ_V and θ_I are their phase angles. From here on, let us assume that the reference phase angle of the voltage source θ_V, is zero, and let $\theta_I = \theta$.

Since the instantaneous power by a circuit element is given by the product of the instantaneous voltage and current, it is possible to obtain a general expression for the power dissipated by an AC circuit element:

$$p(t) = v(t)i(t) = V_m I_m \cos(\omega t)\cos(\omega t + \theta) \tag{6.1.2}$$

Equation (6. 1. 2) can be further simplified as:

$$p(t) = \frac{V_m I_m}{2}\cos(\theta) + \frac{V_m I_m}{2}\cos(2\omega t + \theta) \tag{6.1.3}$$

where θ is the difference in phase between voltage and current. Equation (6. 1. 3) illustrates that the instantaneous power dissipated by an AC circuit element is equal to the sum of an average component, $\frac{1}{2}VI\cos(\theta)$, plus a sinusoidal component, $\frac{VI}{2}\cos(2\omega t + \theta)$, oscillating at a frequency double that of the original source frequency.

6.1.1 Power for a purely resistive load

First, consider the case in which the network is a pure resistance. Then, $\theta = 0$, and we have

$$\begin{aligned} v(t) &= V_m \cos(\omega t) \\ i(t) &= I_m \cos(\omega t) \\ p(t) &= v(t)i(t) = V_m I_m \cos^2(\omega t) \end{aligned} \tag{6.1.4}$$

Plots of these quantities are shown in Figure 6. 1. 1. Notice that the current is in phase with the voltage. Because $p(t)$ is positive at all times, we conclude that energy flows continually in the direction from the source to the load (where it is converted to heat). The value of the power rises and falls with the voltage (and current) magnitude.

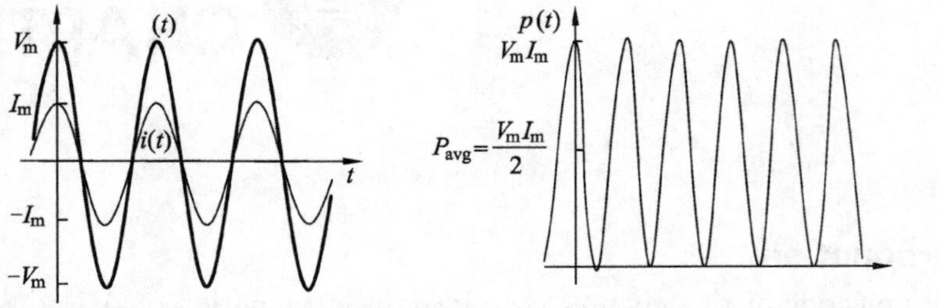

Figure 6.1.1 Current voltage and power versus time for a purely resistive load

6.1.2 Power for a pure inductive load

Consider the case in which the load is a pure inductance for which $Z=\omega L\angle 90°$. So, $\theta=90°$, and we get

$$\begin{aligned} v(t)&=V_m\cos(\omega t)\\ i(t)&=I_m\cos(\omega t-90°)=I_m\sin(\omega t)\\ p(t)&=v(t)i(t)=V_mI_m\cos(\omega t)\sin(\omega t)\\ &=\frac{V_mI_m}{2}\sin(2\omega t)\end{aligned} \tag{6.1.5}$$

Plots of the current, voltage, and power are shown in Figure 6.1.2(a). Notice that the current lags the voltage by 90°. For half of the time the power is positive, showing that energy is delivered to the inductance, where it is stored in the magnetic field. For the other half of the time, power is negative, showing that the inductance returns energy to the source. Notice that the average power is zero. In this case, we say that reactive power flows from the source to the load.

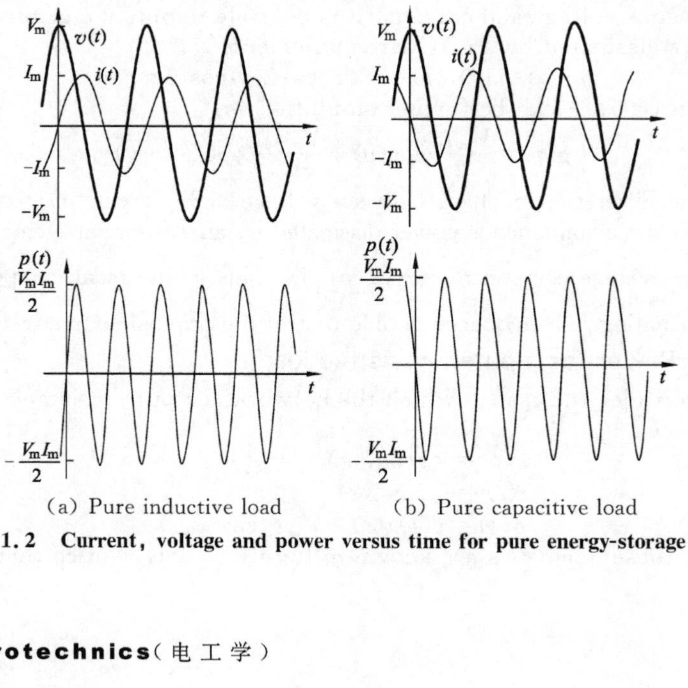

Figure 6.1.2 Current, voltage and power versus time for pure energy-storage elements

6.1.3 Power for a pure capacitive load

Consider the case in which the load is a pure capacitance for which $Z=\frac{1}{\omega C}\angle -90°$. So, $\theta=-90°$, and we get

$$\begin{aligned} v(t) &= V_m\cos(\omega t) \\ i(t) &= I_m\cos(\omega t+90°)=-I_m\sin(\omega t) \\ p(t) &= v(t)i(t)=-V_m I_m\cos(\omega t)\sin(\omega t) \\ &= -\frac{V_m I_m}{2}\sin(2\omega t) \end{aligned} \tag{6.1.6}$$

Plots of the current, voltage, and power are shown in Figure 6.1.2(b). Here again, the average power is zero, and we say that reactive power flows. However, the power for the capacitance carries the opposite sign as that for the inductance. So, we say that reactive power is positive for an inductance and is negative for a capacitance. If a load contains both inductance and capacitance with reactive powers of equal magnitude, the reactive powers cancel.

6.2 Average power

The **average power** corresponding to the voltage and current signals of Equation 6.1.1 can be obtained by integrating the instantaneous power over one cycle of the sinusoidal signal. Let $T=2\pi/\omega$ represent one cycle of the sinusoidal signals. Then the **average power** P_{av}, is given by the integral of the instantaneous power, $p(t)$, over one cycle:

$$\begin{aligned} P_{av} &= \frac{1}{T}\int_0^T p(t)\,dt \\ &= \frac{1}{T}\int_0^T \frac{V_m I_m}{2}\cos(\theta)\,dt+\frac{1}{T}\int_0^T \frac{V_m I_m}{2}\cos(2\omega t+\theta)\,dt \end{aligned} \tag{6.2.1}$$

$$P_{av}=\frac{V_m I_m}{2}\cos(\theta) \tag{6.2.2}$$

In phasor notation, the current and voltage of Equation (6.1.1) are given by

$$\begin{aligned} V_m(j\omega) &= V_m e^{j\theta_V} \\ I_m(j\omega) &= I_m e^{j\theta_I} \end{aligned} \tag{6.2.3}$$

Note further that if the impedance of an AC circuit is defined by the phasor voltage and current of Equation (6.2.3) to be

$$Z=\frac{V_m}{I_m}e^{j(\theta_V-\theta_I)}=|Z|e^{j\theta_z} \tag{6.2.4}$$

and therefore, the phase angle of the impedance is $\theta_Z=\theta_V-\theta_I$.

The expression for the average power obtained in Equation (6.2.1) can therefore also be represented using phasor notation, as follows:

$$P_{av}=\frac{1}{2}\frac{V_m^2}{|Z|}\cos\theta_Z=\frac{1}{2}I_m^2|Z|\cos\theta_Z \tag{6.2.5}$$

where V_m and I_m are the peak amplitudes of the sinusoidal voltage and current.

6.3 AC Power Notation

It is customary in AC power analysis to employ the rms value of the AC voltages and currents in the circuit. Thus, the following expressions will be used in this chapter

$$
\begin{aligned}
V_{rms} &= \frac{V}{\sqrt{2}} = \widetilde{V} \\
I_{rms} &= \frac{I}{\sqrt{2}} = \widetilde{I}
\end{aligned} \tag{6.3.1}
$$

$$
\begin{aligned}
P_{av} &= \frac{1}{2}\frac{V^2}{|Z|}\cos\theta_Z = \frac{\widetilde{V}^2}{|Z|}\cos\theta_Z \\
&= \frac{1}{2}I^2|Z|\cos\theta_Z = \widetilde{I}^2|Z|\cos\theta_Z = \widetilde{V}\widetilde{I}\cos\theta_Z
\end{aligned} \tag{6.3.2}
$$

where $\widetilde{V}$ and $\widetilde{I}$ are the rms value of the sinusoidal voltage and current. As usual, the units of average power are watts (W).

The amplitudes of phasor voltages and currents will be denoted throughout this chapter by means of the rms amplitude. We'll use a slight modification in the phasor notation to define the following **rms phasor** quantities:

$$
\begin{aligned}
\widetilde{\mathbf{V}} &= V_{rms}e^{j\theta_V} = \widetilde{V}e^{j\theta_V} = \widetilde{V}\angle\theta_V \\
\widetilde{\mathbf{I}} &= I_{rms}e^{j\theta_I} = \widetilde{I}e^{j\theta_I} = \widetilde{I}\angle\theta_I
\end{aligned} \tag{6.3.3}
$$

In other words, throughout the remainder of this chapter the symbols $\widetilde{V}$ and $\widetilde{I}$ will denote the rms value of a voltage or a current, and the symbols $\widetilde{\mathbf{V}}$ and $\widetilde{\mathbf{I}}$ will denote rms phasor voltages and currents. Thus, the sinusoidal waveform corresponding to the phasor current $\mathbf{I} = \widetilde{I}\angle\theta_I$ corresponds to the time-domain waveform

$$i(t) = \sqrt{2}\widetilde{I}\cos(\omega t + \theta_I) \tag{6.3.4}$$

and the sinusoidal form of the phasor voltage is

$$v(t) = \sqrt{2}\widetilde{V}\cos(\omega t + \theta_V) \tag{6.3.5}$$

Example 6.3.1 Compute average AC power

$v(t) = 14.14\cos(377t)$ V; $R = 4$; $L = 8$ mH. Find: P_{av} for the RL load of Figure for example 6.3.1.

Solution: First, we define the phasors and impedances at the frequency of interest in the problem, $\omega = 377$ rad/s:

$\widetilde{\mathbf{V}} = 10\angle(0°)$ V $Z = R + j\omega L = 4 + j3 = 5\angle(37°)\,\Omega$

$$\widetilde{\mathbf{I}} = \frac{\widetilde{\mathbf{V}}}{Z} = \frac{10\angle(0°)}{5\angle(37°)} = 2\angle(-37°)\text{ A}$$

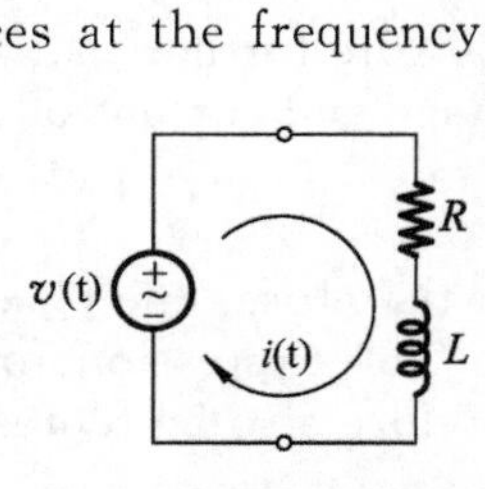

Figure for example 6.3.1

The average power can be computed from the phasor quantities:

$P_{av} = \widetilde{V}\widetilde{I}\cos\theta = 10 \times 2 \times \cos(37°) = 16$ W

Comments: Please pay attention to the use of rms values in this example. It is very important to remember that we have defined phasors to have rms amplitude in power calculation. This is a standard procedure in electrical engineering practice.

Example 6.3.2

Compute the average power dissipated by the load of Figure for example 6.3.2. $\tilde{V}_S = 110\angle 0$ V; $R_S = 2\Omega$; $R_L = 16\Omega$; $C = 100\ \mu$F. Find: P_{av} for the RC Load.

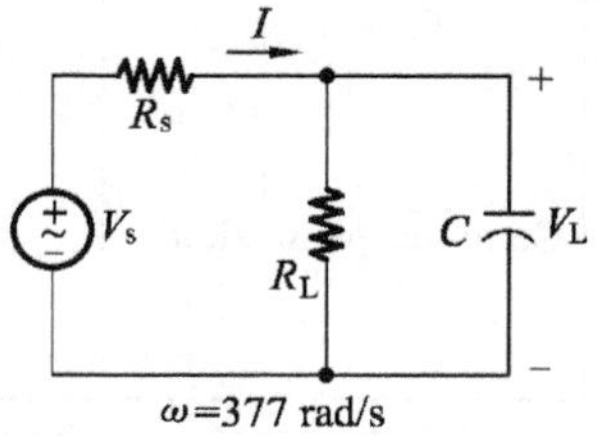

Figure for example 6.3.2

Solution: First, we compute the load impedance at the frequency of interest in the problem, $\omega=377$ rad/s:

$$Z_L = R \parallel \frac{1}{j\omega C} = \frac{R_L}{1+j\omega CR_L}$$

$$= \frac{16}{1+j0.6032} = 13.7\angle(-0.543)\Omega$$

Next, we compute the load voltage, using the voltage divider rule:

$$\mathbf{V}_L = \frac{Z_L}{R_S+Z_L}\tilde{\mathbf{V}}_S$$

$$= \frac{13.7\angle(-0.543)}{2+13.7\angle(-0.543)} 110\angle(0) = 97.6\angle(-0.067)\text{V}$$

Knowing the load voltage, we can compute the average power according to:

$$P_{av} = \frac{|\tilde{\mathbf{V}}_L|^2}{|Z_L|}\cos(\theta) = \frac{97.6^2}{13.7}\cos(-0.543) = 595\ \text{W}$$

or, alternatively, we can compute the load current and calculate average power according to the equation below:

$$\tilde{\mathbf{I}}_L = \frac{\tilde{\mathbf{V}}_L}{Z_L} = 7.1\angle(0.476)\text{A}$$

$$P_{av} = |\tilde{\mathbf{I}}_L|^2 |Z_L| \cos(\theta) = 7.1^2 \times 13.7 \times \cos(-0.543) = 595\ \text{W}$$

6.4 Power factor

The phase angle of the load impedance plays a very important role in the absorption of power by a load impedance. As illustrated in Equation 6.3.2, the average power dissipated by an AC load is dependent on the cosine of the angle of the impedance. To recognize the importance of this factor in AC power computations, the term cos (θ) is referred to as the **power factor** (**pf**). Sometimes, θ is called the power angle.

$$\theta = \theta_V - \theta_I \tag{6.4.1}$$

Note that the power factor is equal to 0 for a pure inductive or capacitive load and equal to 1 for a pure resistive load; in every other case,

$$0 < pf < 1 \tag{6.4.2}$$

Often, power factor is stated as a percentage. Also, it is common to state whether the current leads (capacitive load) or lags (inductive load) the voltage. A typical power factor would be stated to be 90% lagging, which means that $\cos(\theta) = 0.9$ and that the current lags the voltage.

Table 6.4.1 illustrates the concept and summarizes all of the important points so far. In the table, the phasor voltage $\tilde{\mathbf{V}}$ has a zero phase angle and the current phasor is referenced to the phase of $\tilde{\mathbf{V}}$.

Table 6.4.1 Important facts related to complex power

	Resistive load	Capacitive load	Inducitive load
Ohm's law	$\tilde{\boldsymbol{V}}_L = Z_L \tilde{\boldsymbol{I}}_L$	$\tilde{\boldsymbol{V}}_L = Z_L \tilde{\boldsymbol{I}}_L$	$\tilde{\boldsymbol{V}}_L = Z_L \tilde{\boldsymbol{I}}_L$
Complex impedance	$Z_L R_L$	$Z_L = R_L - jX_L$	$Z_L = R_L + jX_L$
Phase angle	$\theta = \theta$	$\theta < \theta$	$\theta > \theta$
Complex plane sketch	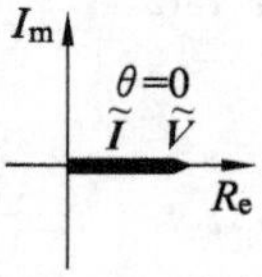	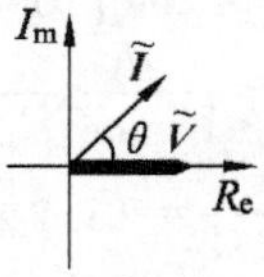	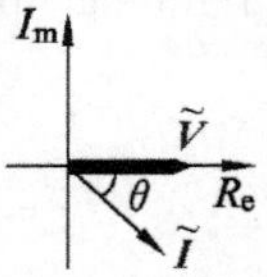
Explanation	The current is in phase with the voltage	The current "leads" the voltage	The current "lags" the voltage
Power factor	Umity	Leading, <1	Lagging, <1
Reactive power	0	Negative	Positive

6.5 Reactive power

In AC circuits, energy flows into and out of energy storage elements (inductances and capacitances). For example, when the voltage magnitude across a capacitance is increasing, energy flows into it, and when the voltage magnitude decreases, energy flows out. Similarly, energy flows into an inductance when the current flowing through it increases in magnitude. Although instantaneous power can be very large, the net energy transferred per cycle is zero for either an ideal capacitance or inductance.

When a capacitance and an inductance are in parallel (or series), energy flows into one, while it flows out of the other. Thus, the power flow of a capacitance tends to cancel that of an inductance at each instant in time.

The peak instantaneous power associated with the energy storage elements contained in a general load is called reactive power and is given by

$$Q = \tilde{V}\tilde{I}\sin(\theta) \tag{6.5.1}$$

where θ is the power angle given by Equation (6.4.1). (Notice that, to a pure resistance, $\theta = 0$ and $Q = 0$.)

The physical units of reactive power are watts. However, to emphasize the fact that Q doesn't represent the flow of net energy, its units are usually given as Volt Amperes Reactive (VAR).

Since P_{av} corresponds to the power absorbed by the load resistance, it is also called the **real power**, measured in units of watts (W). On the other hand, Q takes the name of **reactive power**, since it is associated with the load reactance. Table 6.1.2 shows the general methods of calculating P and Q.

where $\qquad R = |Z|\cos\theta,\ X = |Z|\sin\theta$

are the resistive and reactive components of the load impedance, respectively.

Note that since reactive elements can only store energy and not dissipate it, there is no net average power absorbed by X.

Table 6.1.2 Real and Reactive power

Real power, P_{av}	Reactive power, Q
$\tilde{V}\tilde{I}\cos(\theta)$	$\tilde{V}\tilde{I}\sin(\theta)$
$\tilde{I}^2 R$	$\tilde{I}^2 X$

6.6 Complex power

The computation of AC power is greatly simplified by defining a fictitious but very useful quantity called the **complex power**, S:

$$S=\tilde{V}\tilde{I}^* \tag{6.6.1}$$

where the asterisk denotes the complex conjugate. You may easily verify this definition to the convenient expression

$$S=\tilde{V}\tilde{I}\cos\theta+\mathrm{j}\tilde{V}\tilde{I}\sin\theta \tag{6.6.2}$$

or

$$S=P_{av}+\mathrm{j}Q \tag{6.6.3}$$

The magnitude of S and $|S|$, is measured in units of **volt-amperes (VA)** and is called **apparent power.** Because this is the quantity that would be computed by measuring the rms load voltage and currents without regard for the phase angle of the load.

If we know the complex power S, then we can find the average power, reactive power and apparent power:

$$\begin{aligned} P&=\mathrm{Re}(S)=\mathrm{Re}\left(\frac{1}{2}\boldsymbol{VI}^*\right)\\ Q&=Im(S)=Im\left(\frac{1}{2}\boldsymbol{VI}^*\right)\\ \text{apparent power}&=|S|=\left|\frac{1}{2}\boldsymbol{VI}^*\right|\end{aligned} \tag{6.6.4}$$

where Re(S) denotes the real part of S and $Im(S)$ denotes the imaginary part of S.

6.7 Units

Often, the units given for a quantity indicate whether the quantity is average power (W), reactive power (VAR), or apparent power (VA). For example, if we have a 5 kW load, this means that $P=5$ kW. On the other hand, if we have a 5 kVA load, this means $\tilde{V}\tilde{I}=5$ kVA. If we say that a load absorbs 5 kVAR, then $Q=5$ kVAR.

Example 6.7.1 Complex power calculations

$v(t)=100\cos(\omega t+15^\circ)$ V; $i(t)=2\cos(\omega t-15^\circ)$ A. Find: $S=P_{av}+\mathrm{j}Q$ for the complex load in Figure for example 6.7.1.

Solution: First, we convert the voltage and current into phasor quantities:

$$\tilde{\boldsymbol{V}}=\frac{100}{\sqrt{2}}\angle(15^\circ)\mathrm{V},\quad \tilde{\boldsymbol{I}}=\frac{2}{\sqrt{2}}\angle(-15^\circ)\mathrm{A}$$

Next, we compute real and reactive power:

$$P_{av}=\tilde{V}\tilde{I}\cos(\theta_V-\theta_I)=\frac{200}{2}\cos(30°)=86.6\ \text{W}$$

$$Q=\tilde{V}\tilde{I}\sin(\theta_V-\theta_I)=\frac{200}{2}\sin(30°)=50\ \text{VAR}$$

Now we use Equation 6.6.3 to get: $S=P_{av}+\text{j}Q=86.6+\text{j}50$ W

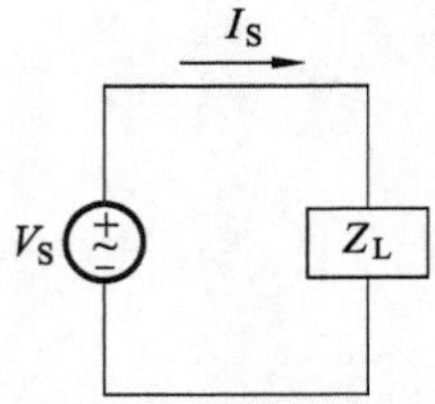

Figure 6.7.1

Example 6.7.2 Real and reactive power calculations

$\tilde{V}=110\angle 0$ V; $R_S=2\ \Omega$; $R_L=5\ \Omega$; $C=2\,000\ \mu$F. Find: Complex power; real and reactive power for the load of Figure for example 6.7.2.

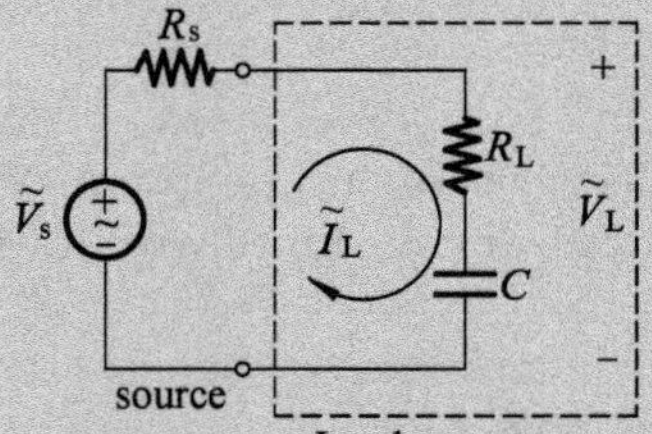

Figure for example 6.7.2

Solution: Define the load impedance:

$$Z_L=R_L+\frac{1}{\text{j}\omega C}=5-\text{j}1.326=5.17\angle(-15°)\Omega$$

Next, compute the load voltage and current:

$$\tilde{\boldsymbol{V}}_L=\frac{Z_L}{R_s+Z_L}\tilde{\boldsymbol{V}}_s=\frac{5-\text{j}1.326}{7-\text{j}1.326}\times 110$$
$$=79.9\angle(-4.13°)\text{V}$$

$$\tilde{I}_L=\frac{\tilde{\boldsymbol{V}}_L}{Z_L}=\frac{79.9\angle(-4.13°)}{5.17\angle(-15°)}$$
$$=15.44\angle(10.87°)\text{A}$$

Finally, we compute the complex power

$$S=\tilde{\boldsymbol{V}}_L\tilde{\boldsymbol{I}}_L^*=79.9\angle(-4.13°)\times 15.44\angle(-10.87°)$$
$$=1233.66\angle(-15°)=1192-\text{j}319\ \text{W}$$

Therefore,

$$P_{av}=1192\ \text{W};\ Q=-319\ \text{VAR}$$

6.8 Power-factor correction

We have seen that large currents can flow in energy-storage devices (inductance and capacitance) without average power being delivered. In heavy industry, many loads are partly inductive, and large amounts of reactive power flow. This reactive power causes higher currents in the power distribution system. Consequently, the lines and transformers must have higher ratings than would be necessary to deliver the same average power to a resistive (100 percent power factor) load. Power-factor correction can provide a significant economic advantage for large consumers of electrical energy.

Energy rates charged to industry depend on the power factor, with higher charges for energy delivered at lower power factors. (Power factor is not taken into account for residential customers). Therefore, it is advantageous to choose loads that operate at near unity power factor.

A common approach is to place capacitors in parallel with an inductive load to increase the power factor.

Example 6.8.1

A 50 kW load operates from 50 Hz 10 kV rms line with a power factor of 60% lagging. Compute the capacitance that must be placed in parallel with the load to achieve a 90% lagging power factor.

Solution: First, we find the load power factor

$$\theta_L=\arccos(0.6)=53°$$

Then the reactive power of the load is

$$Q_L=P_L\tan(\theta_L)=50\times\tan(53°)=66\ \text{kVAR}$$

After adding the capacitor, the power will still be 50 kW and the power angle will become

$$\theta_{new}=\arccos(0.9)=26°$$

The new value of the reactive power will be

$$Q_{new}=P_L\tan(\theta_{new})=50\times\tan(26°)=24\ \text{kVAR}$$

So, the reactive power of the capacitance must be

$$Q_C=Q_{new}-Q_L=-42\ \text{kVAR}$$

Now, we find that the reactance of the capacitor is

$$X_C=\frac{V_{rms}^2}{Q_C}-2356\ \Omega$$

Finally, the angular frequency is

$$\omega=2\pi\times 50=314$$

And the capacitance required is

$$C=\frac{1}{\omega|X_C|}=\frac{1}{314\times 2356}=1.35\times 10^{-6}\text{F}$$

Summary

1. When a sinusoidal current flows through a sinusoidal voltage, the average power delivered is $P=V_{rms}I_{rms}\cos(\theta)$, where θ is the power angle, which is found by subtracting the phase angle of the current from the phase angle of the voltage. The power factor is $\cos(\theta)$.

2. Reactive power is the flow of energy back and forth between the source and energy-storage elements(L and C). We defined reactive power to be positive for an inductance and negative for a capacitance. The net energy transferred per cycle by reactive power flow is zero.

3. Apparent power is the product of rms voltage and rms current. The unit for it is VA.

4. Power factor is the cosine of the angle θ by which the current lags the voltage (If the current leads the voltage, the angle is negative). If the current lags the voltage, the power factor is said to be inductive or lagging. If the current leads the voltage, the power factor is said to be capacitive or leading.

Problems

6.1 Determine the complex power, power, reactive power and apparent power absorbed by the load in Figure for problem 6.1. Also, determine the power factor for the load.

6.2 Compute the average power dissipated by the load of Figure for problem 6.2.

Known quantities: $\tilde{V}_S = 110\angle 0$ V; $R = 10\ \Omega$; $L = 0.05$ H; $C = 470\ \mu$F.

Find: P_{av} for the complex load.

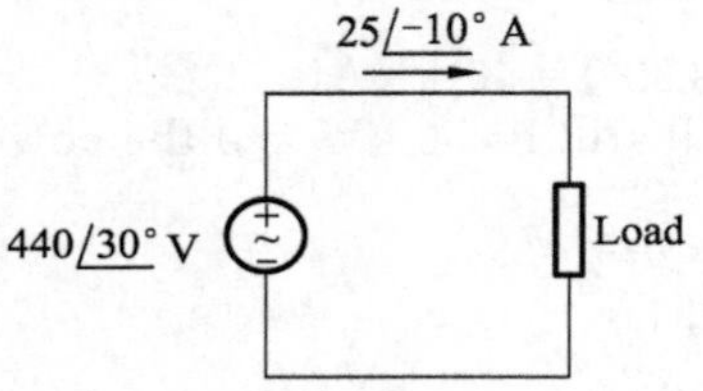

Figure for example 6.1

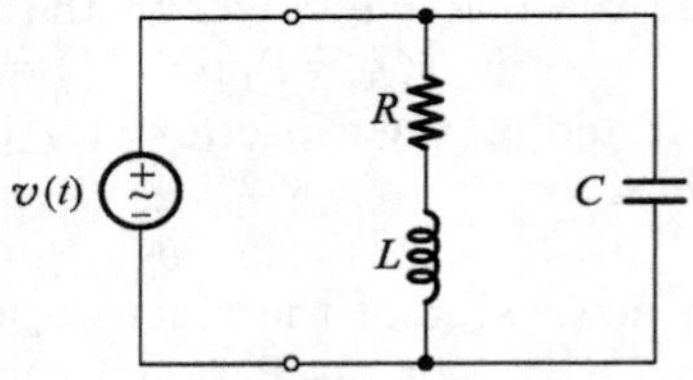

Figure for example 6.2

6.3 Consider the circuit shown in Figure for problem 6.3. Find the load impedance of the circuit, and compute the average power dissipated by the load.

6.4 Compute the power factor for an inductive load with $L = 100$ mH and $R = 0.4\ \Omega$. Assume $\omega = 377$ rad/s.

Find the reactive and real power for the load of Figure for problem 6.4.

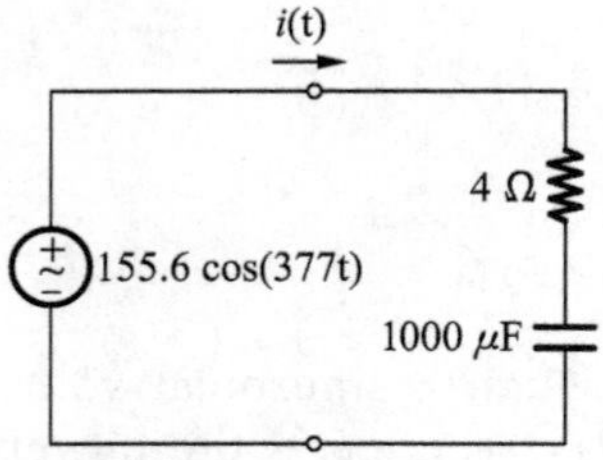

Figure for example 6.3

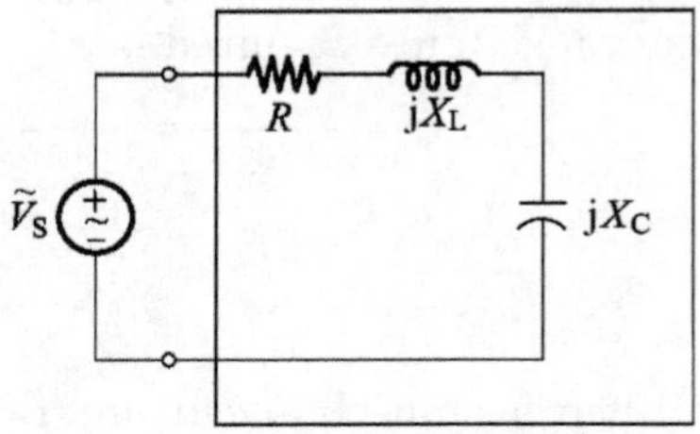

Figure for example 6.4

CHAPTER 7
BANLANCED THREE-PHASE CIRCUITS

Introduction

Most of the AC power used today is generated and distributed as **three-phase power**, by means of an arrangement in which three sinusoidal voltages are generated out of phase with each other. The primary reason is efficiency: the weight of the conductors and other components in a three-phase system is much lower than in a single-phase system delivering the same amount of power. Another advantage of balanced three-phase systems is that the total power is constant (as a function of time) rather than pulsating.

In this chapter we will study wye-connected (Y-connected) source. Load impedances can be either wye connected or delta connected.

7.1 Balanced three-phase source

We consider the most common case: three equal-amplitude AC voltages having phases that are 120° apart. This is known as a balanced three-phase source.

The three voltages shown in Figure 7.1.1(a) are given by

$$\begin{aligned} v_{an}(t) &= \sqrt{2}\tilde{V}_{an}\cos(\omega t) \\ v_{bn}(t) &= \sqrt{2}\tilde{V}_{bn}\cos(\omega t - 120^\circ) \\ v_{cn}(t) &= \sqrt{2}\tilde{V}_{cn}\cos(\omega t + 120^\circ) \end{aligned} \tag{7.1.1}$$

where the quantities $\tilde{V}_{an}$, $\tilde{V}_{bn}$ and $\tilde{V}_{cn}$ are rms values of three-phase voltages and are equal to each other. To simplify the notation, it will be assumed from here on that

$$\tilde{V}_{an} = \tilde{V}_{bn} = \tilde{V}_{cn} = \tilde{V}_P \tag{7.1.2}$$

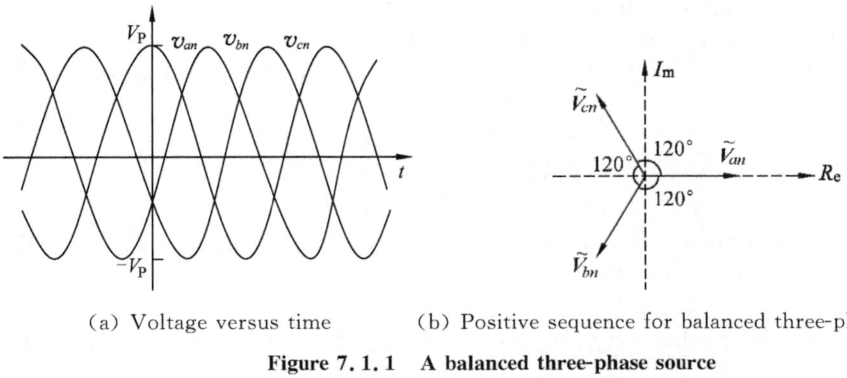

(a) Voltage versus time (b) Positive sequence for balanced three-phase voltages

Figure 7.1.1 A balanced three-phase source

The corresponding phasors are

$$\begin{aligned}\tilde{\boldsymbol{V}}_{an}&=\tilde{V}_{P}\angle 0^{\circ}\\ \tilde{\boldsymbol{V}}_{bn}&=\tilde{V}_{P}\angle -120^{\circ}\\ \tilde{\boldsymbol{V}}_{cn}&=\tilde{V}_{P}\angle -240^{\circ}=\tilde{V}_{P}\angle +120^{\circ}\end{aligned}\tag{7.1.3}$$

The phasor diagram is shown in Figure 7. 1. 1(b).

Obviously $$v_{an}(t)+v_{bn}(t)+v_{cn}(t)=0\tag{7.1.4}$$

or $$\tilde{\boldsymbol{V}}_{an}+\tilde{\boldsymbol{V}}_{bn}+\tilde{\boldsymbol{V}}_{cn}=0\tag{7.1.5}$$

This kind of voltages are called banlanced three-phase source.

7.1.1 Phase sequence

Three-phase sources can have either a positive or negative phase sequence. Referring to Figure 7. 1. 1(b) we notice that v_{an} leads v_{bn}, which in turn leads v_{cn}. We think that the phasors are rotating counterclockwise in determining phase relationships. This set of voltages is said to have a **positive phase sequence** because the voltages reach their peak values in the order $a\rightarrow b\rightarrow c$. If we interchanged b and c, we would have a **negative phase sequence**, in which the order is $a\rightarrow c\rightarrow b$.

Phase sequence can be important. For example, if we have a three-phase induction motor, the direction of rotation is opposite to the two-phase sequences. It can be reversed by changing the phase sequence. To reverse the direction of rotation of such a motor, for example, we would just interchange the b and c connections.

Because circuit analysis is very similar for both phase sequences, we consider only the positive phase sequence in most of the discussion below.

7.1.2 Wye-(or-Y)connected three-phase source

The source shown in Figure 7. 1. 2 is called wye connected (Y-connected) three-phase source. The wire a, b, c are called lines, and the wire n is called neutral. This configuration is called a three-phase four-wire system.

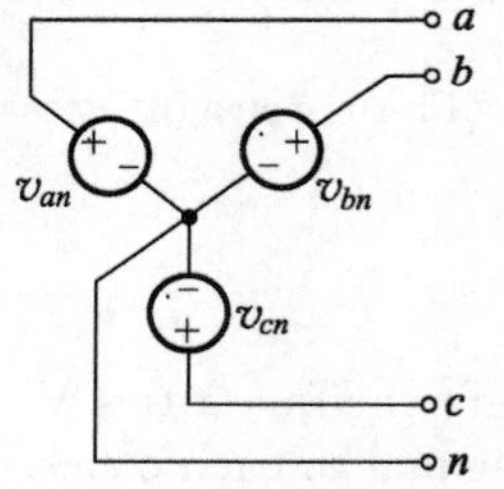

Figure 7. 1. 2 Y-connected three-phase source

The voltages between terminals a, b, or c and the neutral point n are called **line-to-neutral voltages or phase voltages.** On the other hand, voltages between a and b, b and c, or a and c are called **line-to-line voltages** or, more simply, **line voltages.** Thus, v_{an}, v_{bn}, and v_{cn} are **phase voltages**, whereas v_{ab}, v_{bc}, and v_{ca} are **line voltages.** Let us consider the relationships between them.

We can obtain the following relationship by applying KVL to Figure 7. 1. 2:

$$\begin{aligned}\tilde{\boldsymbol{V}}_{ab}&=\tilde{\boldsymbol{V}}_{an}-\tilde{\boldsymbol{V}}_{bn}\\ &=\tilde{V}_{P}\angle 0^{\circ}-\tilde{V}_{P}\angle -120^{\circ}\\ &=\sqrt{3}\tilde{V}_{P}\angle 30^{\circ}\end{aligned}\tag{7.1.6}$$

We denote the magnitude of the line voltages as $\tilde{V}_{L}$.

It can be seen, then, that the magnitude of the line voltages is $\sqrt{3}$ times the magnitude of the phase voltages:

$$\tilde{V}_{L}=\sqrt{3}\tilde{V}_{P}\tag{7.1.7}$$

Thus, the relationship between the line voltage $\tilde{\boldsymbol{V}}_{ab}$ and the phase voltage $\tilde{\boldsymbol{V}}_{an}$

is

$$\tilde{\mathbf{V}}_{ab}=\tilde{\mathbf{V}}_{an}\times\sqrt{3}\angle 30^\circ \quad (7.1.8)$$

Similarly, it can be shown that

$$\tilde{\mathbf{V}}_{bc}=\tilde{\mathbf{V}}_{bn}\times\sqrt{3}\angle 30^\circ \quad (7.1.9)$$

and

$$\tilde{\mathbf{V}}_{ca}=\tilde{\mathbf{V}}_{cn}\times\sqrt{3}\angle 30^\circ \quad (7.1.10)$$

These voltages are shown in Figure 7.1.4.

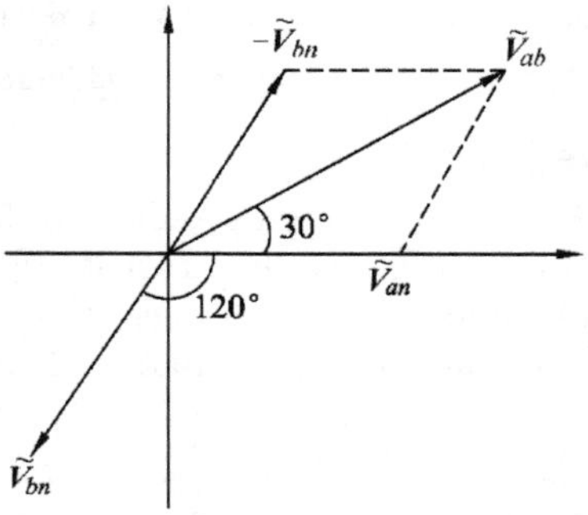

Figure 7.1.3 Phasor diagram showing the relationship between the line-to-line voltage and the-line-to-neutral voltage

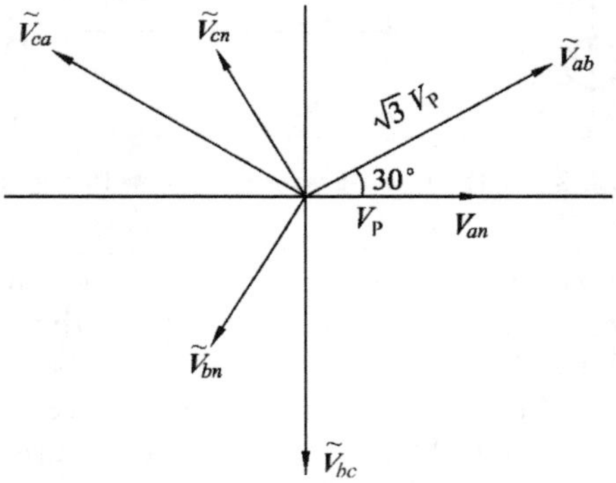

Figure 7.1.4 All phasors starting from the origin

7.1.3 Delta-connected three-phase source

A set of balanced three-phase voltage sources can also be connected in the form of a delta, as shown in Figure 7.1.5. Ordinarily, we avoid connecting voltage sources in closed loops. However, in this case, it turns out that the sum of the voltage is zero:

$$\tilde{\mathbf{V}}_{ab}+\tilde{\mathbf{V}}_{bc}+\tilde{\mathbf{V}}_{ca}=0 \quad (7.1.11)$$

Thus, the current circulating in the delta connected is zero. Clearly, a delta-connected source has no neutral point, so only a three-wire connection is possible for it.

7.2 Load impedances in three-phase circuits

Load impedances can be either wye-connected or delta-connected, as shown in Figure 7. 2. 1.

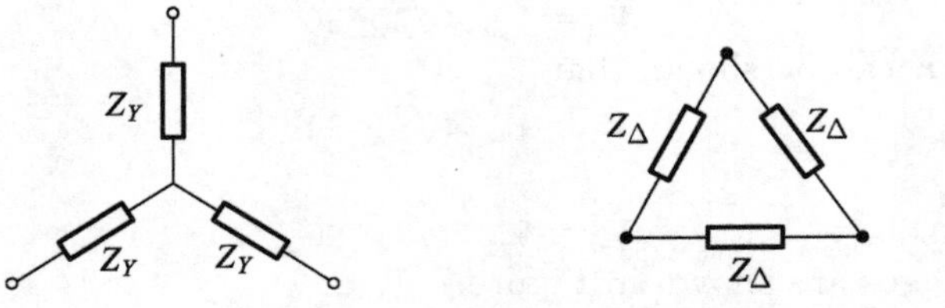

(a) Wye-connected load (b) Delta-connected load

Figure 7. 2. 1 The connection of load impedances

7.2.1 Balanced wye loads

Consider the three-phase source connected to a balanced three-phase load shown in Figure 7. 2. 2. This configuration is called a wye-wye (Y-Y) connection with neutral. By the term balanced load, we mean that the three load impedances are equal, (In this book, we consider only balanced loads.)

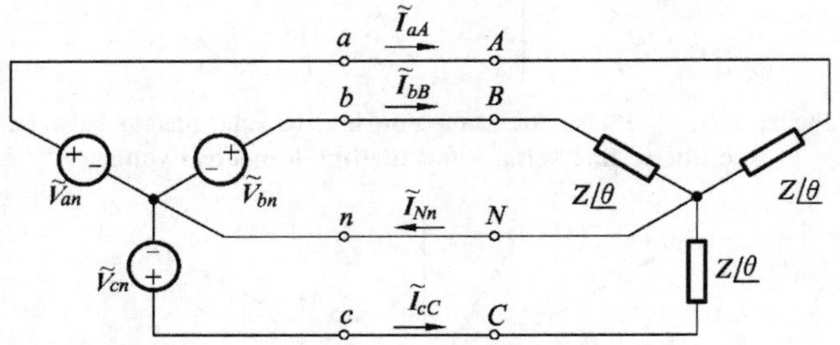

Figure 7. 2. 2 A three-phase wye-wye connection with neutral

$$\boldsymbol{Z}_a = \boldsymbol{Z}_b = \boldsymbol{Z}_c = \boldsymbol{Z}_Y = |Z_Y| \angle \theta \quad (7.2.1)$$

We often use the term phase to refer to part of the source or the load. Thus, phase A of the source is $v_{an}(t)$, and phase A of the load is the impedance connected between A and N. the current $\tilde{\boldsymbol{I}}_{AN}$ is called phase current. Furthermore, $\tilde{\boldsymbol{I}}_{aA}$, $\tilde{\boldsymbol{I}}_{bB}$, and $\tilde{\boldsymbol{I}}_{cC}$ are called line currents. $\tilde{\boldsymbol{I}}_{aA}$ is the current referenced from node a to node A, as illustrated in Figure 7. 2. 2.

From the diagram of Figure 7. 2. 2, it can be verified that each impedance gets the corresponding phase voltage across itself, The current in phase A of the load is given by

$$\tilde{\boldsymbol{I}}_{aA} = \frac{\tilde{\boldsymbol{V}}_{an}}{Z\angle\theta} = \frac{\tilde{V}_{P}\angle 0^{\circ}}{Z\angle\theta} = \tilde{I}_{L}\angle -\theta \quad (7.2.2)$$

where $I_L = V_P / Z$ is the rms value of the line current. Because the load impedances are equal, all of the line currents are the same, except for phase. The phase angles of the currents will differ by $\pm 120^{\circ}$. Thus, the currents are given by

$$\begin{aligned} i_{aA}(t) &= \sqrt{2}\tilde{I}_L \cos(\omega t - \theta) \\ i_{bB}(t) &= \sqrt{2}\tilde{I}_L \cos(\omega t - \theta - 120^{\circ}) \\ i_{cC}(t) &= \sqrt{2}\tilde{I}_L \cos(\omega t - \theta + 120^{\circ}) \end{aligned} \quad (7.2.3)$$

The neutral current in Figure 7. 2. 2 is given by

$$i_{Nn}(t) = i_{aA}(t) + i_{bB}(t) + i_{cC}(t) \quad (7.2.4)$$

In terms of phasors, this is

$$\begin{aligned}\tilde{\boldsymbol{I}}_{Nn} &= \tilde{\boldsymbol{I}}_{aA}+\tilde{\boldsymbol{I}}_{bB}+\tilde{\boldsymbol{I}}_{cC} \\ &= \tilde{I}_L\angle-\theta+\tilde{I}_L\angle(-\theta-120^\circ)+\tilde{I}_L\angle(-\theta+120^\circ) \\ &= 0\end{aligned} \tag{7.2.5}$$

Thus the sum of three phasors with equal magnitudes and 120° apart in phase is zero.

It's shown that the neutral current is zero in a balanced three-phase system. Consequently, the neutral wire can be eliminated without changing any of the voltages or currents. Then, the three source voltages are delivered to the three load impedances with three wires.

7.2.2 Wye-delta connection

In addition to a wye connection, it is also possible to connect a balanced load in the delta configuration. A wye-connected generator and a delta-connected load are shown in Figure 7. 2. 3. The load impedances are connected between lines. Thus, the voltage of each load impedance equals to line voltages. We assume that the source voltages are given by

$$\begin{aligned}\tilde{\boldsymbol{V}}_{ab} &= \tilde{V}_L\angle 30^\circ \\ \tilde{\boldsymbol{V}}_{bc} &= \tilde{V}_L\angle -90^\circ \\ \tilde{\boldsymbol{V}}_{ca} &= \tilde{V}_L\angle 150^\circ\end{aligned} \tag{7.2.6}$$

and the impedance of each phase of the load is $Z_\Delta\angle\theta$.

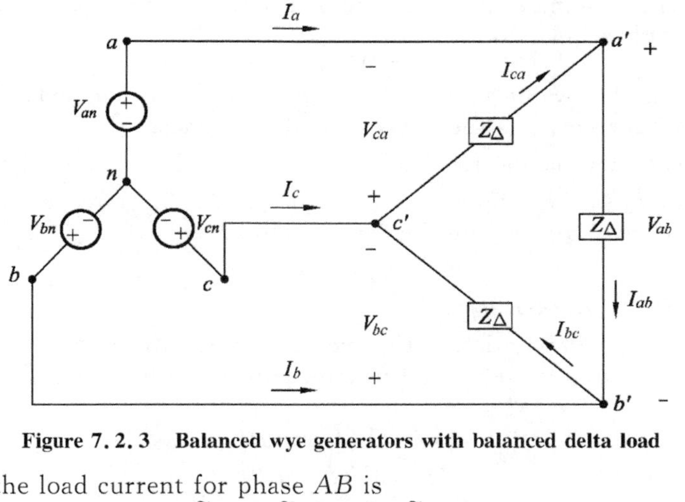

Figure 7. 2. 3 Balanced wye generators with balanced delta load

Then, the load current for phase AB is

$$\tilde{\boldsymbol{I}}_{AB}=\frac{\tilde{\boldsymbol{V}}_{ab}}{Z_\Delta\angle\theta}=\frac{\tilde{V}_L\angle 30^\circ}{Z_\Delta\angle\theta}=\frac{\tilde{V}_L}{Z_\Delta}\angle(30^\circ-\theta)$$

We define the magnitude of the current as

$$\tilde{I}_\Delta=\frac{\tilde{V}_L}{Z_\Delta} \tag{7.2.7}$$

Then

$$\tilde{\boldsymbol{I}}_{AB}=\tilde{I}_\Delta\angle(30^\circ-\theta) \tag{7.2.8}$$

Similarly,

$$\tilde{\boldsymbol{I}}_{BC}=\tilde{I}_\Delta\angle(-90^\circ-\theta) \tag{7.2.9}$$

$$\tilde{\boldsymbol{I}}_{CA}=\tilde{I}_\Delta\angle(150^\circ-\theta) \tag{7.2.10}$$

The current in line a-A is

$$\begin{aligned}\tilde{\boldsymbol{I}}_{aA}&=\tilde{\boldsymbol{I}}_{ab}-\tilde{\boldsymbol{I}}_{ca}\\&=\tilde{I}_{\Delta}\angle(30^{\circ}-\theta)-\tilde{I}_{\Delta}\angle(150^{\circ}-\theta)\\&=\tilde{\boldsymbol{I}}_{ab}\times\sqrt{3}\angle-30^{\circ}\end{aligned}\tag{7.2.11}$$

So the magnitude of the line current is

$$\tilde{I}_{L}=\sqrt{3}\tilde{I}_{\Delta}\tag{7.2.12}$$

It shows that for a balanced delta-connected load, the magnitude of the line-current is equal to $\sqrt{3}$ times the current magnitude in any arm of the delta-connected.

7.3 Three-phase power

7.3.1 Average power

It is possible to compute the power for each phase by considering the phase voltages (equal to the load voltage) for each impedance, and the associated load current. Thus:

$$\begin{aligned}P_{A}(t)&=\tilde{V}_{an}\tilde{I}_{a}\cos(\theta_{a})\\P_{B}(t)&=\tilde{V}_{bn}\tilde{I}_{b}\cos(\theta_{b})\\P_{C}(t)&=\tilde{V}_{cn}\tilde{I}_{c}\cos(\theta_{c})\end{aligned}\tag{7.3.1}$$

Thus, the total average power is

$$P_{av}=P_{a}+P_{b}+P_{c}\tag{7.3.2}$$

Because the load impedances are equal, $\tilde{\boldsymbol{I}}_{a}$, $\tilde{\boldsymbol{I}}_{b}$ and $\tilde{\boldsymbol{I}}_{c}$ have the same rms value $\tilde{I}_{P}$, except for phase. Thus,

$$P_{av}=3\tilde{V}_{P}\tilde{I}_{P}\cos(\theta)\tag{7.3.3}$$

where $\tilde{V}_{P}$ is the rms value of the phase voltages of each impedance; $\tilde{I}_{P}$ is the associated load current; θ is the impedance of each phase.

To balanced wye-connected loads, $\tilde{V}_{L}=\sqrt{3}\tilde{V}_{P}$, $\tilde{I}_{L}=\tilde{I}_{P}$; to delta-connected loads $\tilde{V}_{L}=\tilde{V}_{P}$, $\tilde{I}_{L}=\sqrt{3}\tilde{I}_{P}$. Substitute to Equation (7.3.3) respectively, and we can get another way to compute the average power:

$$P_{av}=\sqrt{3}\tilde{V}_{L}\tilde{I}_{L}\cos(\theta)\tag{7.3.4}$$

7.3.2 Reactive power

Similar to the computation of the average power, the reactive power of each impedance are equal, too. The total reactive power of a balanced three-phase load is given by

$$Q=3\tilde{V}_{P}\tilde{I}_{P}\sin(\theta)\tag{7.3.5}$$

or

$$Q=\sqrt{3}\tilde{V}_{L}\tilde{I}_{L}\sin(\theta)\tag{7.3.6}$$

7.3.3 Complex power

Let us denote the complex power for each phase by $\boldsymbol{S}$:

$$\begin{aligned}\boldsymbol{S}&=\tilde{\boldsymbol{V}}\cdot\tilde{\boldsymbol{I}}^{*}\\&=\tilde{V}_{P}\tilde{I}_{P}\cos(\theta)+\mathrm{j}\tilde{V}_{P}\tilde{I}_{P}\sin(\theta)\\&=P+\mathrm{j}Q\end{aligned}\tag{7.3.7}$$

where $\tilde{V}_{P}$ and $\tilde{I}_{P}$ denote, once again, the rms values of each phase voltage and phase current. Consequently, the total real power delivered to the balanced wye load is $3P$, and the total reactive power is $3Q$. Thus, the total complex power, S_{T}, is given by

$$S_T = P_T + jQ_T = 3P + j3Q \tag{7.3.8}$$

and the apparent power is

$$\begin{aligned}|\boldsymbol{S}_T| &= \sqrt{(3P^2)+(3Q^2)} \\ &= 3\sqrt{(\tilde{V}_P\tilde{I}_P\cos\theta)^2+(\tilde{V}_P\tilde{I}_P\sin\theta)^2} \\ &= 3\tilde{V}_P\tilde{I}_P\end{aligned} \tag{7.3.9}$$

or

$$|\boldsymbol{S}_T| = \sqrt{3}\tilde{V}_L\tilde{I}_L \tag{7.3.10}$$

7.3.4 Instantaneous power of the balanced three-phase circuit

The voltages of the three-balanced source are given by

$$v_{an}(t)=\sqrt{2}\tilde{V}_P\cos(\omega t)$$
$$v_{bn}(t)=\sqrt{2}\tilde{V}_P\cos(\omega t-120^\circ)$$
$$v_{cn}(t)=\sqrt{2}\tilde{V}_P\cos(\omega t+120^\circ)$$

The currents of the balanced three-phase loads are given by

$$i_a(t)=\sqrt{2}\tilde{I}_P\cos(\omega t)$$
$$i_b(t)=\sqrt{2}\tilde{I}_P\cos(\omega t-120^\circ)$$
$$i_c(t)=\sqrt{2}\tilde{I}_P\cos(\omega t+120^\circ)$$

We obtain

$$\begin{aligned}p(t)=&\sqrt{2}\tilde{V}_P\cos(\omega t)\sqrt{2}\tilde{I}_P\cos(\omega t-\theta)+\\ &\sqrt{2}\tilde{V}_P\cos(\omega t-120^\circ)\sqrt{2}\tilde{I}_P\cos(\omega t-\theta-120^\circ)+\\ &\sqrt{2}\tilde{V}_P\cos(\omega t+120^\circ)\sqrt{2}\tilde{I}_P\cos(\omega t-\theta)\end{aligned}$$

Using the trigonometric identity

$$\cos(x)\cos(y)=\frac{1}{2}\cos(x-y)+\frac{1}{2}\cos(x+y)$$

we find that

$$p(t)=3\tilde{V}_P\tilde{I}_P\cos(\theta)+\tilde{V}_P\tilde{I}_P[\cos(2\omega t-\theta)+\cos(2\omega t-\theta-240^\circ)+\cos(2\omega t-\theta-480^\circ)]$$

the term in the bracket

$$\cos(2\omega t-\theta)+\cos(2\omega t-\theta-240^\circ)+\cos(2\omega t-\theta+480^\circ)=0$$

So the power becomes

$$p(t)=3\tilde{V}_P\tilde{I}_P\cos(\theta)=\textbf{a constant!} \tag{7.3.11}$$

Notice that the total power is constant with respect to time. It's an important characteristic of a balanced three-phase power system.

A consequence of this fact is that the torque required to drive a three-phase generator connected to a balanced load is constant, and vibration is lessened. Similarly, the torque produced by a three-phase motor is constant rather than pulsating as it is for a single-phase motor.

Example 7.3.1 Analysis of a Wye-Wye system

A balanced positive-sequence wye-connected 60 Hz three-phase source has line-to-neutral voltages of $\widetilde{V}_P = 1\,000$ V. This source is connected to a balanced wye-connected load. Each phase of the load consists of a 0.1 H inductance in series with a 50 Ω resistance. Find the line currents, the line-to line voltages, the power, and the reactive power delivered to the load. Assume that the phase angle of $\widetilde{\boldsymbol{V}}_{an}$ is zero.

Solution:

$$Z = R + \mathrm{j}X_L = 50 + \mathrm{j}37.70 = 62.62\angle 37°\ \Omega$$

Draw the circuit as shown in Figure for example 7. 3. 1. In balanced wye-wye-calculations, we can assume that n and N are connceted. Thus,

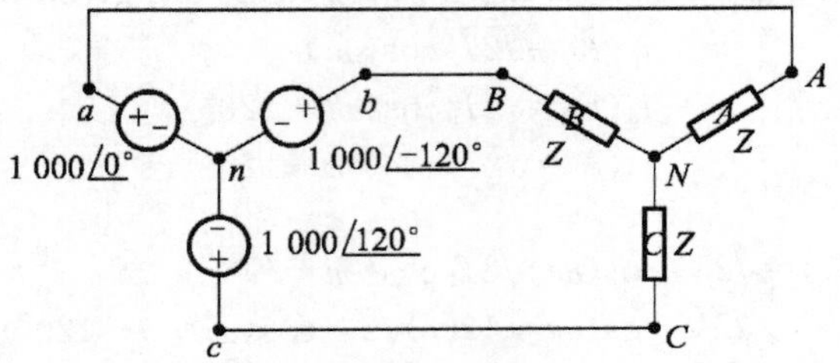

Figure for example 7. 3. 1

$$\widetilde{\boldsymbol{I}}_{aA} = \frac{\widetilde{\boldsymbol{V}}_{an}}{Z} = \frac{1\,000\angle 0°}{62.62\angle 37°} = 15.97\angle -37°\ \mathrm{A}$$

Similarly,

$$\widetilde{\boldsymbol{I}}_{bB} = \frac{\widetilde{\boldsymbol{V}}_{bn}}{Z} = \frac{1\,000\angle -120°}{62.62\angle 37°} = 15.97\angle -157°\ \mathrm{A}$$

$$\widetilde{\boldsymbol{I}}_{cC} = \frac{\widetilde{\boldsymbol{V}}_{cn}}{Z} = \frac{1\,000\angle 120°}{62.62\angle 37°} = 15.97\angle 83°\ \mathrm{A}$$

The line-to-line voltages are

$$\widetilde{\boldsymbol{V}}_{ab} = \widetilde{\boldsymbol{V}}_{an} \times \sqrt{3}\angle 30° = 1\,732\angle 30°\ \mathrm{V}$$

$$\widetilde{\boldsymbol{V}}_{bc} = \widetilde{\boldsymbol{V}}_{bn} \times \sqrt{3}\angle 30° = 1\,732\angle -90°\ \mathrm{V}$$

$$\widetilde{\boldsymbol{V}}_{ca} = \widetilde{\boldsymbol{V}}_{cn} \times \sqrt{3}\angle 30° = 1\,732\angle 150°\ \mathrm{V}$$

The power delivered to the load is

$$P = 3\widetilde{V}_P\widetilde{I}_P\cos(\theta) = 3 \times 1\,000 \times 15.97 \times \cos(37°) = 38.26\ \mathrm{kW}$$

The reactive power is

$$Q = 3\widetilde{V}_P\widetilde{I}_P\sin(\theta) = 3 \times 1\,000 \times 15.97 \times \sin(37°) = 28.84\ \mathrm{kVAR}$$

Summary

Electric power is most commonly generated in three-phase form, for reasons of efficiency. Three-phase power entails the generation of three 120° out-of-phase AC voltages of equal amplitude, so that the instantaneous power is actually constant.

Three-phase sources and loads can be configured in either wye or delta configurations; of these, the wye form is more common. The calculation of currents, voltages, and power in three-phase circuits is greatly simplified if one uses per-phase calculations.

Problems

7.1 Compute the power delivered to the balanced load of Figure for problem 7.1 if the lines have zero resistance and $Z_L = 1 + j3$.

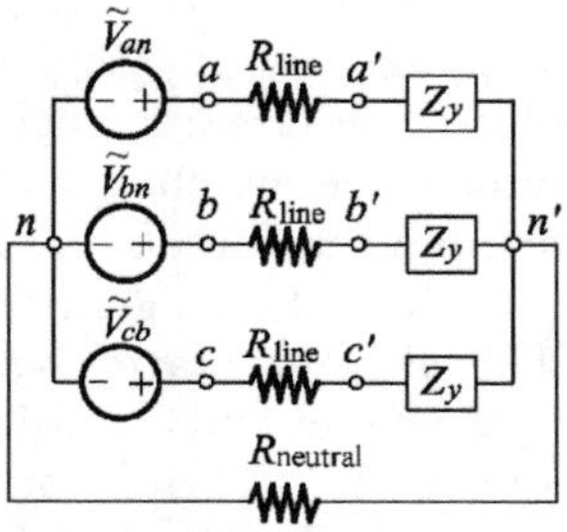

Figure for problem 7.1

7.2 The phase currents in a four-wire wye-connected load are as follows: $\tilde{I}_{an} = 10\angle 0$ A, $\tilde{I}_{bn} = 12\angle 150°$ A, $\tilde{I}_{cn} = 8\angle 1.09°$ A. Determine the current in the neutral wire.

7.3 For the circuit shown in Figure for example 7.3, find the currents $\tilde{I}_R$, $\tilde{I}_W$, $\tilde{I}_B$ and $\tilde{I}_N$.

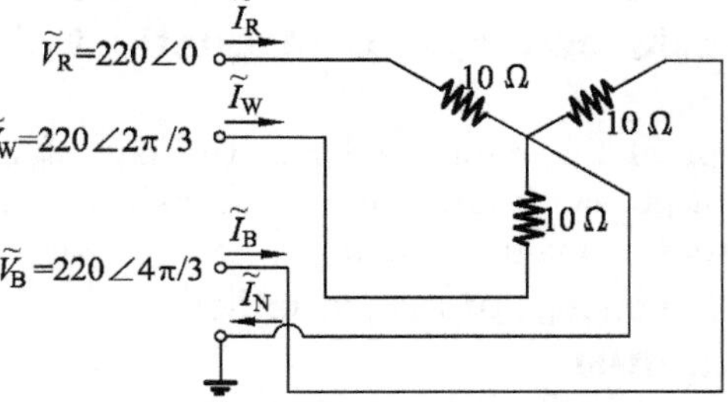

Figure for problem 7.3

7.4 For the circuit shown in Figure for exmple 7.4, find the currents $\tilde{I}_A$, $\tilde{I}_B$, $\tilde{I}_C$, and $\tilde{I}_N$, and the real power dissipated by the load.

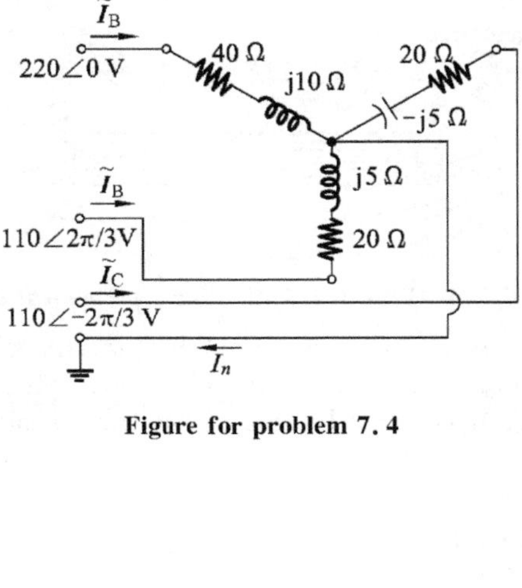

Figure for problem 7.4

CHAPTER 8
TRANSFORMERS

Introduction

Transformers are key circuit elements. They are used in power systems for stepping up or stepping down AC voltages or currents. They are used in electronic circuits such as radio and television receivers for such purposes as impedance matching, isolating one part of a circuit from another, and again for stepping up or down AC voltages and currents. The aim of this chapter is to introduce the magnetic circuit concepts, and the simple transformers calculation. The chapter starts with the basic concept of magnetic field and the law of magnetic circuit, and it serves as the foundation to the transformers. The next subject is a brief introduction to the characteristics and types of magnetic material. Then, the emphasis moves to the introduction of transformers. We will consider the linear transformers and the ideal transformers.

8.1 The basic concept of magnetic field and the law of magnetic circuit

The basic concept of magnetic field and the law of magnetic circuit is the basis for learning the basic knowledge of transformers. Let's review some important characteristics of magnetic field.

8.1.1 The basic concept of magnetic field

8.1.1.1 Magnetic field

In the region surrounding a permanent magnet there exists a magnetic field, which can be represented by magnetic flux lines similar to electric flux lines. Magnetic flux lines, however, do not have origins or terminating points as electric flux lines do but exist in continuous loops, as shown in Figure 8.1.1.

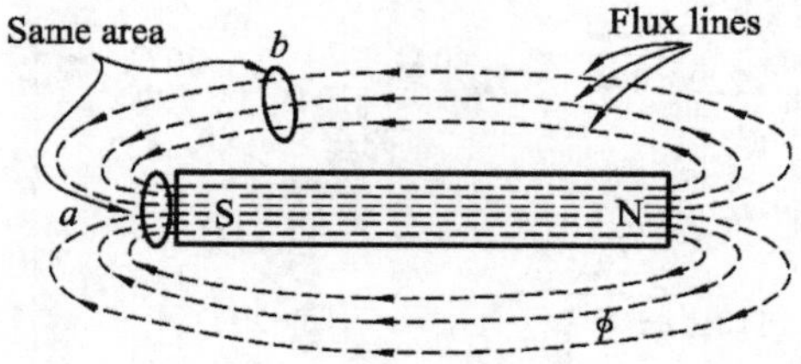

Figure 8.1.1 Flux distribution for a permanent magnet

The magnetic flux lines radiate from the north pole to the south pole, returning to the north pole through the metallic bar. The strength of a magnetic field in a particular region is directly related to the density of flux lines in that region. In Figure 8. 1. 1, for example, the magnetic field strength at a is twice that at b since twice as many magnetic flux lines are associated with the perpendicular plane at a than at b.

If unlike poles of two permanent magnets are brought together, the magnets will attract, and the flux distribution will be as shown in Figure 8. 1. 2. If like poles are brought together, the magnets will repel, and the flux distribution will be as shown in Figure 8. 1. 3.

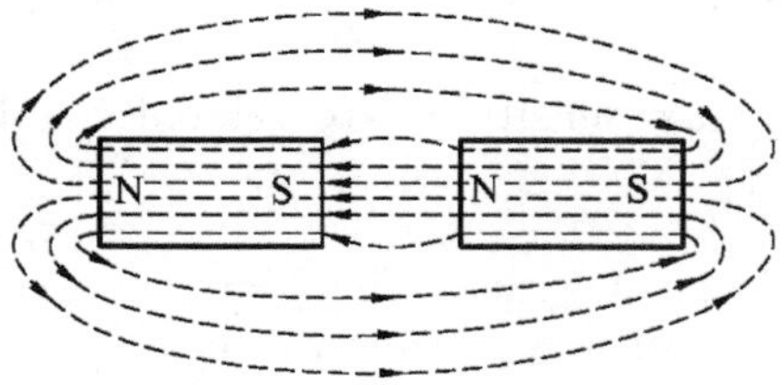

Figure 8. 1. 2 Flux distribution for two adjacent, opposite poles

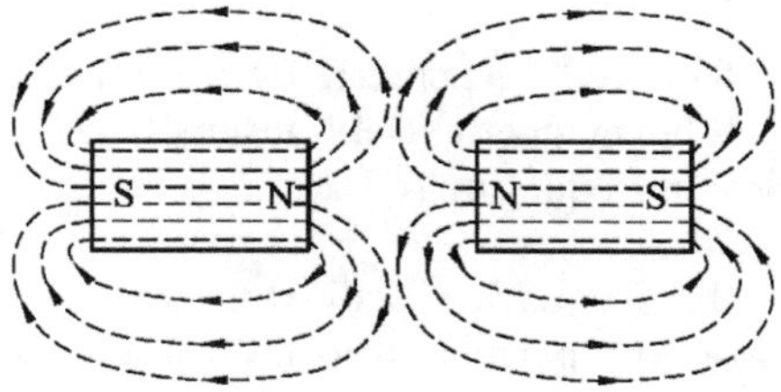

Figure 8. 1. 3 Flux distribution for two adjacent, like poles

8. 1. 1. 2 Magnetic flux

The quantities used to quantify the strength of a magnetic field are the **magnetic flux** ϕ, in units of webers (Wb); and the **magnetic flux density** B, in units of webers per square meter (Wb/m^2), or teslas (T).

The **magnetic flux** ϕ is defined as the integral of the flux density over some surface area. For the simplified case of magnetic flux lines perpendicular to a cross-sectional area equation reference goes here A, we can see that the flux is given by the integral

$$\phi = \int_A B \mathrm{d}A \tag{8. 1. 1}$$

in webers, where the subscript A indicated that the integral is evaluated over surface A. Furthermore, if the flux were to be uniform over the cross-sectional area A, the preceding integral could be approximated by the following expression:

$$\phi = B \cdot A \tag{8. 1. 2}$$

Figure 8. 1. 4 illustrates this idea, by showing hypothetical magnetic flux lines traversing a surface, delimited in the figure by a thin conducting wire.

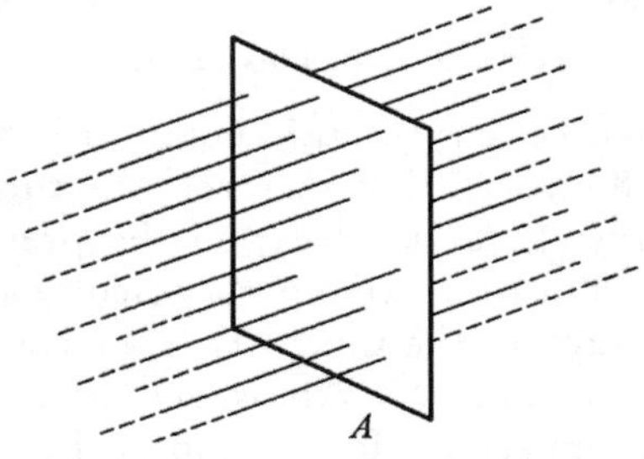

Figure 8. 1. 4 Magnetic flux lines crossing a surface

8. 1. 1. 3 Permeability

If cores of different materials with the same physical dimensions are used in the electromagnet, the strength of the magnet will vary in accordance with the core used. This variation in strength is due to the greater or lesser number of flux lines passing through the core. Materials in which flux lines can readily be set up are said to be magnetic and to have high permeability. The **permeability** (μ) of a material, therefore, is a measure of the case with which magnetic flux lines can be established in the material. It is similar in many respects to conductivity in electric circuits. The permeability of free space μ_0 (Vacuum) is

$$\mu_0 = 4\pi \times 10^{-7}\ \text{Wb/A} \cdot \text{m} \tag{8.1.3}$$

8.1.1.4 Magnetic field intensity

The **magnetic field intensity** H and the flux density B are related by the following equation:

$$B = \mu H \tag{8.1.4}$$

This equation indicates that for a particular magnetic field intensity, the greater the permeability is, the greater the induced flux density will be.

8.1.2 The law of magnetic circuit

8.1.2.1 Faraday's law

Faraday's law states that if the imaginary surface A were bounded by a conductor (for example, the thin wire of Figure 8.1.4), then a changing magnetic field would induce a voltage, and therefore a current, in the conductor. More precisely, Faraday's law states that a time-varying flux causes an induced electromotive force, or **emf**, e, as follows:

$$e = -\frac{d\phi}{dt} \tag{8.1.5}$$

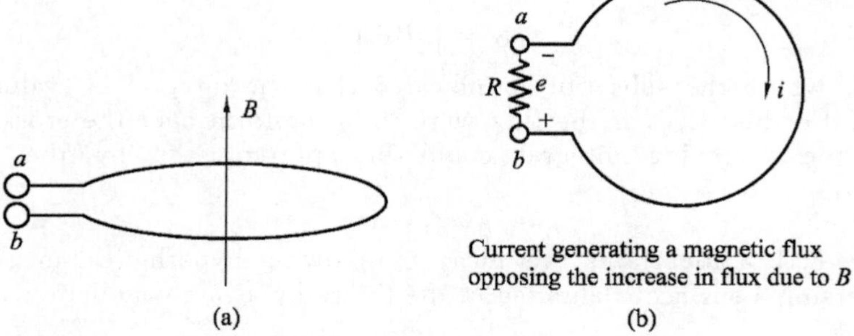

Figure 8.1.5 Flux direction

A little discussion is necessary at this point to explain the meaning of the minus sign in Equation (8.1.5). Consider the one-turn coil of Figure 8.1.5, which forms a circular cross-sectional area, in the presence of a magnetic field with flux density B oriented in a direction perpendicular to the plane of the coil. If the magnetic field, and therefore the flux within the coil, is constant, no voltage will exist across terminals a and b; if, however, the flux were increasing and terminals a and b were connected—for example, by means of a resistor, as indicated in Figure 8.1.5 (b)—current would flow in the coil in such a way that the magnetic flux generated by the current would oppose the increasing flux. Thus, the flux induced by such a current would be in the direction opposite to that of the original flux density vector, B. This principle is known as **Lenz's law.** The reaction flux would then point downward in Figure 8. 1. 5 (a), or into the page in Figure 8. 1. 5 (b). Now, by virtue of the **right-hand rule**, this reaction flux would induce a current flowing clockwise in Figure 8. 1. 5 (b), that is, a current that flows out of terminal b and into terminal a. The resulting voltage across the hypothetical resistor R would then be negative. If, on the other hand, the original flux were decreasing, current would be induced in the coil so as to reestablish the initial flux; but this would mean that the current would have to generate a flux in the upward direction in Figure 8. 1. 5 (a) [or out of the page in Figure 8. 1. 5 (b)]. Thus, the resulting voltage would change sign. Current generating a magnetic flux opposing the increase in flux due to B. The polarity of the induced voltage can usually be determined from physical considerations; therefore, the minus sign in Equation (8. 1. 5) is usually left out.

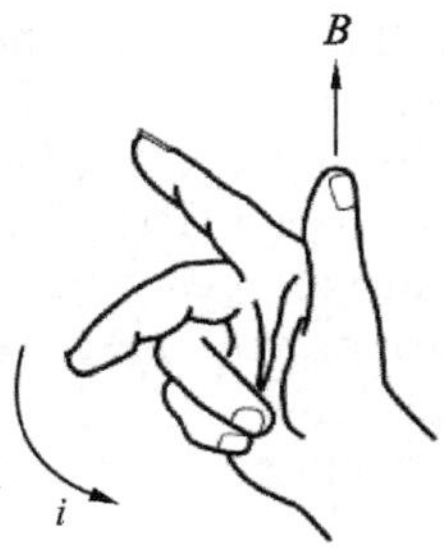

(a) Right-hand rule

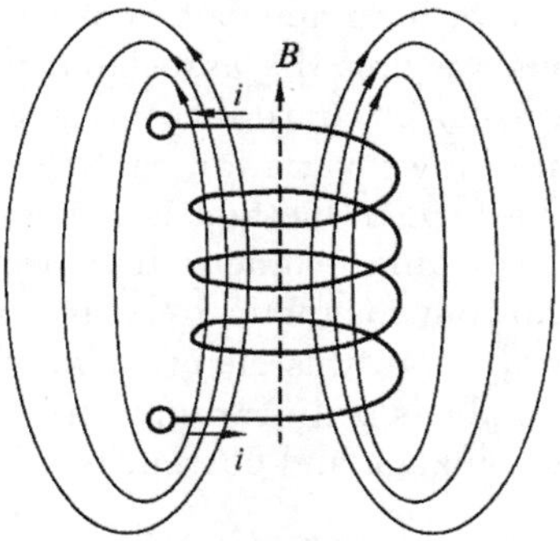

(b) Flux lines

Figure 8. 1. 6 Concept of flux linkage

In practical applications, the size of the voltages induced by the changing magnetic field can be significantly increased if the conducting wire is coiled many times around, so as to multiply the area crossed by the magnetic flux lines many times over. For an N turn coil with cross-sectional area A, for example, we have the emf

$$e=N\frac{\mathrm{d}\phi}{\mathrm{d}t} \tag{8. 1. 6}$$

Figure 8. 1. 6 shows an N turn coil linking a certain amount of magnetic flux; you can see that if N is very large and the coil is tightly wound (as is usually the case in the construction of practical devices), it is not unreasonable to presume that each turn of the coil links the same flux. It is convenient, in practice, to define the flux linkage, λ, as

$$\lambda=N\phi \tag{8. 1. 7}$$

So that

$$e=\frac{\mathrm{d}\lambda}{\mathrm{d}t} \tag{8. 1. 8}$$

Note that Equation (8. 1. 8), relating the derivative of the flux linkage to the induced emf, is analogous to the equation describing current as the derivative of charge:

$$i=\frac{dq}{dt} \tag{8.1.9}$$

In other words, flux linkage can be viewed as the dual of charge in a circuit analysis sense, provided that we are aware of the simplifying assumptions just stated in the preceding paragraphs, namely, a uniform magnetic field perpendicular to the area delimited by a tightly wound coil. These assumptions are not unreasonable at all when applied to the inductor coils commonly employed in electric circuits.

What, then, are the physical mechanisms that can cause magnetic flux to change, and therefore to induce an electromotive force. Two such mechanisms are possible. The first consists of physically moving a permanent magnet in the vicinity of a coil—for example, so as to create a time-varying flux. The second requires that we first produce a magnetic field by means of an electric current (how this can be accomplished is discussed later in this section) and then vary the current, thus varying the associated magnetic field. The latter method is more practical in many circumstances, since it does not require the use of permanent magnets and allows variation of field strength by varying the applied current; however, the former method is conceptually simpler to visualize. The voltages induced by a moving magnetic field are called motional voltages; those generated by a time-varying magnetic field are termed transformer voltages. We shall be interested in both in this chapter, for different applications.

In the analysis of linear circuits, we implicitly assumed that the relationship between flux linkage and current is a linear one:

$$\lambda=Li \tag{8.1.10}$$

so that the effect of a time-varying current is to induce a transformer voltage across an inductor coil, according to the expression

$$v=L\frac{di}{dt} \tag{8.1.11}$$

This is, in fact, the defining equation for the ideal self-inductance, L. In addition to self-inductance, however, it is also important to consider the magnetic coupling that can occur between neighboring circuits. Self-inductance measures the voltage induced in a circuit by the magnetic field generated by a current flowing in the same circuit. It is also possible that a second circuit in the vicinity of the first may experience an induced voltage as a consequence of the magnetic field generated in the first circuit. As we shall see later, this principle underlies the operation of all transformers.

8.1.2.2 Ampère's Law

Ampère's law is one of two fundamental laws relating electricity to magnetism. Ampère's law states that the integral of the vector magnetic field intensity H around a closed path is equal to the total current linked by the closed path i:

$$\oint H \cdot dl = \sum i \tag{8.1.12}$$

where dl is an increment in the direction of the closed path. If the path is in the same direction as the direction of the magnetic field, we can use scalar quantities to state that

$$\int H\mathrm{d}l = \sum i \tag{8.1.13}$$

Figure 8.1.7 illustrates the case of a wire carrying a current i and of a circular path of radius r surrounding the wire. In this simple case, you can see that the magnetic field intensity H is determined by the familiar right-hand rule. This rule states that if the direction of current i points in the direction of the thumb of one's right hand. The resulting magnetic field encircles the conductor in the direction in which the other four lingers would encircle it. Thus, in the case of Figure 8.1.7, the closed-path integral becomes equal to $H \cdot 2\pi r$, since the path and the magnetic field are in the same direction, and therefore the magnitude of the magnetic field intensity is given by

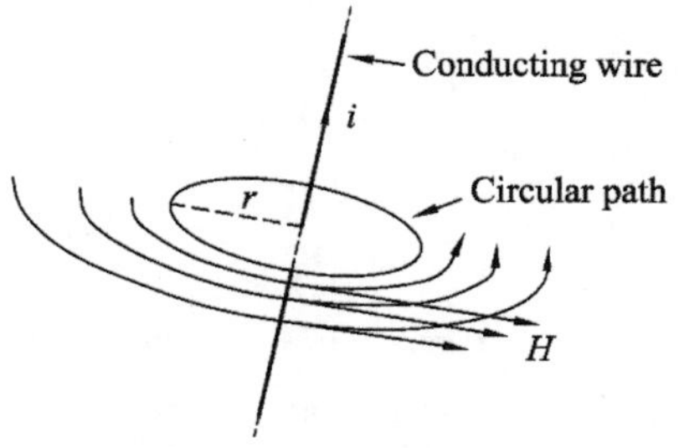

Figure 8.1.7 Illustration of Ampère's law

$$H=\frac{i}{2\pi r} \tag{8.1.14}$$

8.1.2.3 Ohm's law for magnetic circuits

Consider the following relationship:

$$Effect=\frac{Cause}{opposition} \tag{8.1.15}$$

Every conversion of energy to another can be related to this equation. Ohm's law for magnetic circuits is similar to Ohm's law for electric circuits. For magnetic circuits, the effect desired in the flux ϕ. The cause is the **magnetomotive force (mmf)** (F), which is the external force (or "pressure") required to set up the magnetic flux lines within the magnetic material. The opposition setting up of the flux ϕ is the **reluctance** $\mathscr{R}$.

Substituting, we have

$$\phi=\frac{F}{\mathscr{R}} \tag{8.1.16}$$

where $\mathscr{R}$ is the reluctance. The **reluctance** of a material to the setting up of magnetic flux lines in the material is determined by the following equation:

$$\mathscr{R}=\frac{l}{\mu A} \tag{8.1.17}$$

where l is the length of the magnetic path, and A is the cross-sectional area. Note that the reluctance is inversely proportional to the area, indicating that an increase in area will result in a reduction in each and an increase in the desired result: flux. For an increase in length the opposite is true, and the desired effect is reduced. The reluctance, however, is inversely proportional to the permeability. The larger the μ is, the smaller the reluctance will be.

The **magnetomotive farce** F is proportional to the product of the number of turns around the core (in which the flux is to be established) and the current through the turns of wire (Figure 8.1.8). In equation form:

$$F=Ni \tag{8.1.18}$$

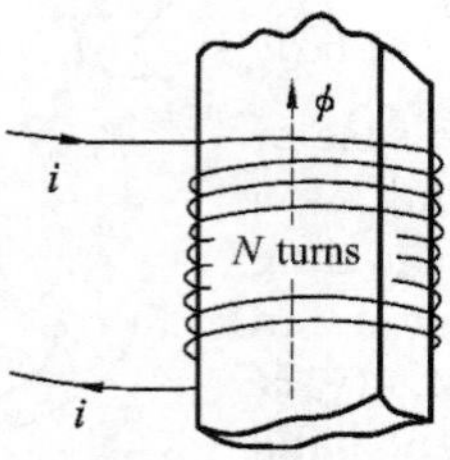

Figure 8.1.8 Defining the components of a magnetiomotive force

This equation clearly indicates that an increase in the number of turns or the current through the wire will result in an increased "pressure" on the system to establish flux lines through the core.

Although there is a great deal of similarity between electric and magnetic circuits, one must continue to realize that the flux ϕ is not a "flow" variable such as current in an electric circuit. Magnetic flux is established in the core through the alteration of the atomic structure of the core due to external pressure and is not a measure of the flow of some charged particles through the core.

8.1.2.4 Kirchhoff's law for magnetic circuits

In electric circuits, we have Kirchhoff's current law and Kirchhoff's voltage law. In magnetic circuits, Kirchhoff's law is also in effect.

When we apply the relationships in magnetic circuits to Kirchhoff's current law, we will find that the sum of the fluxes entering a junction is equal to the sum of the fluxes leaving a junction; that is, for the circuit of Figure 8.1.9,

$$\phi_a = \phi_b + \phi_c \quad \text{(at junction } a\text{)} \tag{8.1.19}$$

or

$$\phi_b + \phi_c = \phi_a \quad \text{(at junction } b\text{)} \tag{8.1.20}$$

both of which are equivalent.

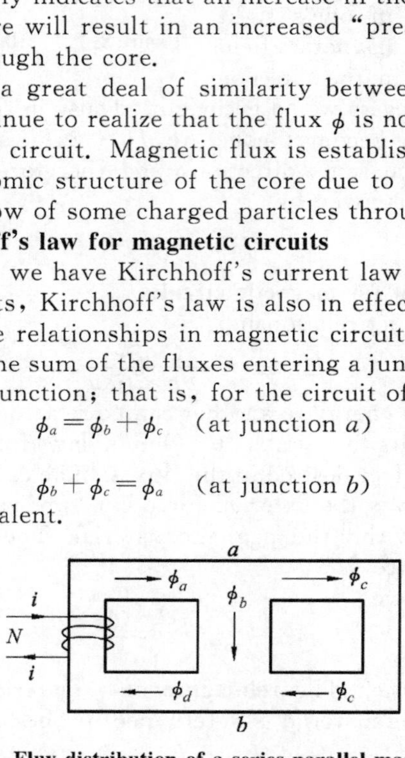

Figure 8.1.9 Flux distribution of a series-parallel magnetic network

If we continue to apply the analogy to Kirchhoff's voltage law, we will obtain the following result:

$$\Sigma_{\zeta} F = 0 \text{ (for magnetic circuits)} \tag{8.1.21}$$

which, in words, states that the algebraic sum of the rises and drops of the mmf around a closed loop of a magnetic circuit is equal to zero; that is, the sum of the rises in mmf equals the sum of the drops in mmf around a closed loop.

The sources of mmf are expressed by the equation

$$F = Ni \tag{8.1.22}$$

The equation for the mmf drop across a portion of a magnetic circuit can be related in the following equation for magnetic circuits:

$$F = \phi \mathcal{R} \tag{8.1.23}$$

where ϕ is the flux passing through a section of the magnetic circuit and $\mathcal{R}$ is the reluctance of that section. The reluctance, however, is seldom calculated in the analysis of magnetic circuits. A more practical equation for the mmf drop is

$$F = Hl(\mathrm{At}) \tag{8.1.24}$$

where H is the **magnetizing force** on a section of a magnetic circuit and l is the length of the section.

As an example of Equation (8.1.21), considering the magnetic circuit appearing in Figure 8.1.10 constructed of three different materials, we have:

$$Ni = \sum_{k=1}^{m} H_k l_k m \tag{8.1.25}$$

where m is the number of the section.

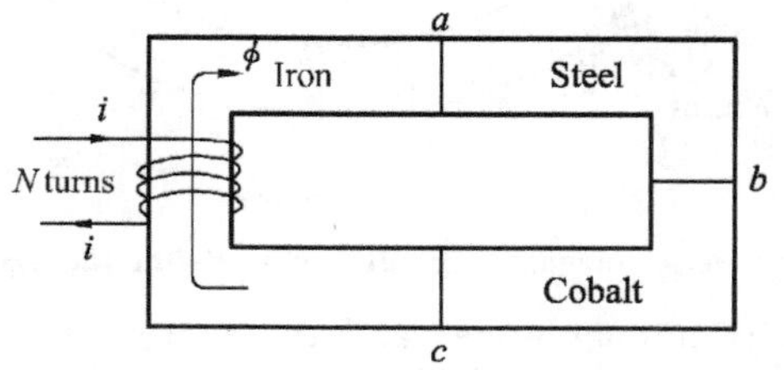

Figure 8.1.10 Series magnetic circuit of three different materials

$$\underbrace{+Ni}_{\text{Rise}} - \underbrace{H_{ab}l_{ab}}_{\text{Drop}} - \underbrace{H_{bc}l_{bc}}_{\text{Drop}} - \underbrace{H_{ca}l_{ca}}_{\text{Drop}} = 0$$
$$\underbrace{Ni}_{\text{Impressed mmf}} = \underbrace{H_{ab}l_{ab} + H_{bc}l_{bc} + H_{ca}l_{ca}}_{\text{mmf drops}} \tag{8.1.26}$$

All the terms of the equation are known except the magnetizing force for each portion of the magnetic circuit, which can be found by using the B-H curve if the flux density B is known.

8.2 Magnetic materials

8.2.1 The classification of magnetic materials

As indicated above, μ has the units of Wb/A • m. Practically speaking, the permeability of all **nonmagnetic materials**, such as copper, aluminum, wood, glass, and air, is the same as that for free space. Materials that have permeability slightly less than that of free space are said to be **diamagnetic**, and those with permeabilities slightly greater than that of free space are said to be **paramagnetic**. Magnetic materials, such as iron, nickel, steel, cobalt, and alloys of these metals, have permeabilities hundreds and even thousands of times that of free space. Materials with these very high permeabilities are referred to as **ferromagnetic**.

The ratio of the permeability of a material to that of free space is called its **relative permeability**; that is,

$$\mu_r = \frac{\mu}{\mu_0} \tag{8.2.1}$$

In general, for ferromagnetic materials, $\mu_r \geqslant 100$, and for nonmagnetic materials, $\mu_r = 1$.

8.2.2 Hysteresis curve

A curve of the flux density B versus the magnetizing force H of a material is of particular importance to the engineer. Curve of this type can usually be found in manuals, descriptive pamphlets, and brochures published by manufacturers of magnetic materials. A typical B-H curve for a ferromagnetic material such as steel can be derived using the setup of Figure 8. 2. 1.

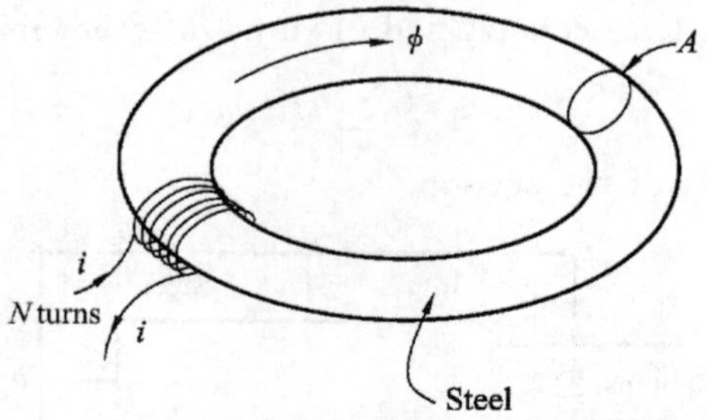

Figure 8. 2. 1 Series magnetic circuit used to define the hysteresis curve

The core is initially unmagnetized and the current $i=0$. If the current i is increased to some value above zero. The magnetizing force H will increase to a value determined by

$$H\uparrow=\frac{Ni\uparrow}{l} \tag{8. 2. 2}$$

The flux ϕ and the flux density $B(B=\phi A)$ will also increase with the current i (or H). If the material has no residual magnetism, and the magnetizing force H is increased from zero to some value H_a, the B-H curve will follow the path shown in Figure 8. 2. 2 between o and a. If the magnetizing force H is increased until saturation H_s occurs, the curve will continue as shown in Figure 8. 2. 2 to point b. When saturation occurs, the flux density has, for all practical purposes, reached its maximum value. Any further increase in current through the coil increasing $H=Ni/l$ will result in a very small increase in flux density B.

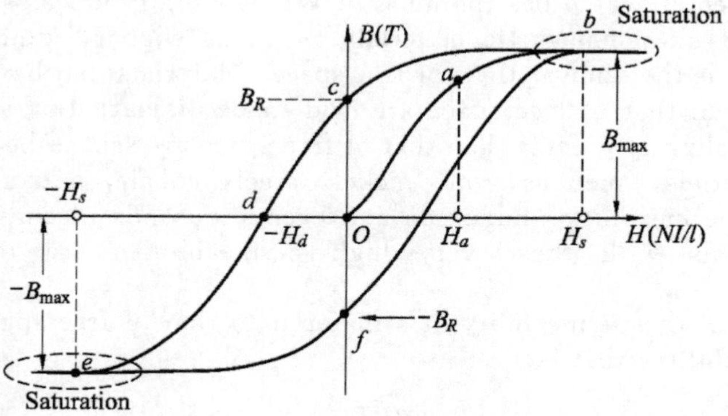

Figure 8. 2. 2 Hysteresis curve

If the magnetizing force is reduced to zero by letting i decrease to zero. The curve will follow the path of the curve between b and c. The flux density B_R, which remains when the magnetizing force is zero—is called the **residual flux density**. It is this residual flux density that makes it possible to create permanent magnets. If the coil is now removed from the core of Figure 8. 2. 1, the core will still have the magnetic properties determined by the residual flux density, a measure of its "retentivity". If the current i is reversed, developing a magnetizing force, $-H$, the flux density B will decrease with an increase in i. Eventually. The flux density will be zero when $-H_d$ (the portion of curve from c to d) is reached. The magnetizing force $-H_d$ required to "coerce" the flux density to reduce its level to zero is called the coercive force, a measure of the coercivity of the magnetic sample. As the force $-H$ is increased until saturation again occurs and is then reversed and brought back to zero, the path *def* will result. If the magnetizing force is increased in the positive direction ($+H$), the curve will trace the path shown from f to b. The entire curve represented by *bcdefb* is called the **hysteresis** curve for the ferromagnetic material, from the Greek hysterein, meaning "to lag behind". The flux density B lagged behind the magnetizing force H during the entire plotting of the curve. When H was zero at c, B was not zero but had only begun to decline. Long after H had passed through zero and had become equal to $-H_d$ did the flux density B finally become equal to zero.

If the entire cycle is repeated, the curve obtained for the same core will be determined by the maximum H applied. Three hysteresis loops for the same materials for maximum values of H less than the saturation value are shown in Figure 8. 2. 3. In addition, the saturation curve is repeated for comparison purposes.

Note from the various curves that for a particular value of H, say H_x, the value of B can vary widely, as determined by the history of the core. In an effort to assign a particular value of B to each value of H, we compromise by connecting the tips of the hysteresis loops. The resulting curve, shown by the heavy, solid line in Figure 8. 2. 3 and for materials in Figure 8. 2. 5, is called the normal magnetization curve. An expanded view of one region appears in Figure 8. 2. 6.

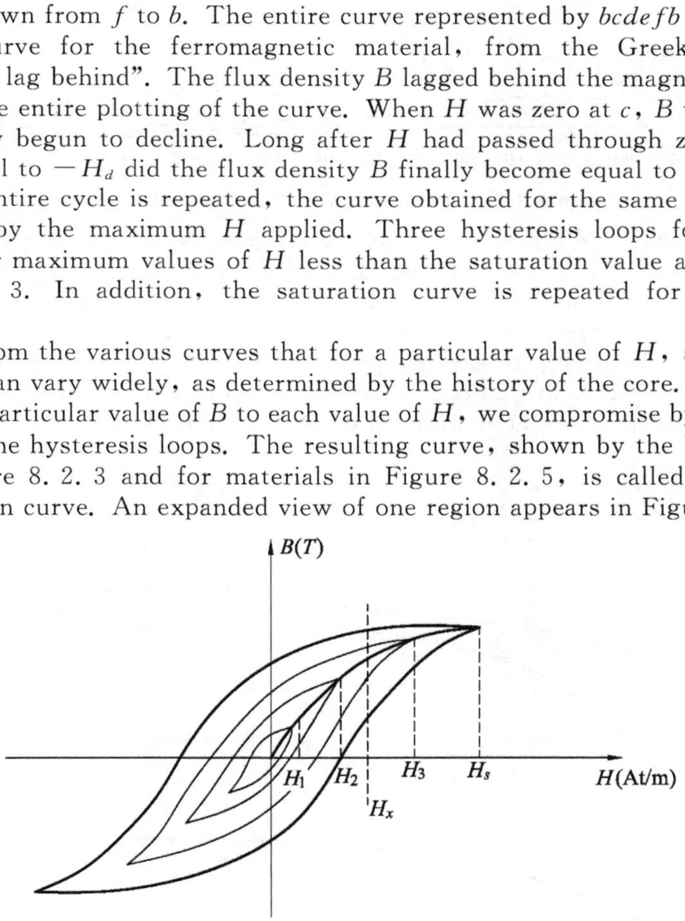

Figure 8. 2. 3 Defining the normal magnetization curve

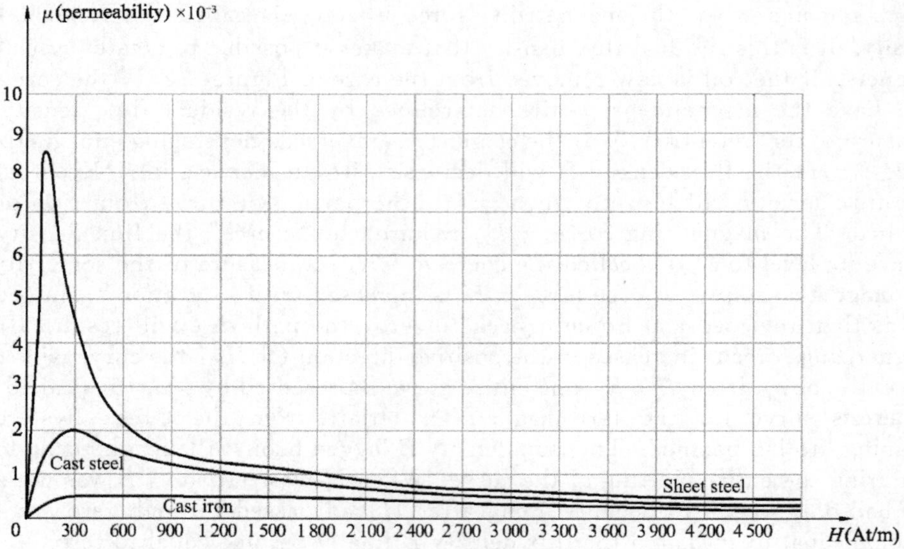

Figure 8. 2. 4 Variation of μ with the magnetizing force

A comparison of Figure 8. 2. 4 and 8. 2. 5 shows that for the same value of H, the value of B is higher in Figure 8. 2. 5 for the materials with the higher μ in Figure 8. 2. 4. This is particularly obvious for low values of H. This correspondence between the two figures must exist since $B=\mu H$. In fact, if in Figure 8. 2. 5, We find μ for each value of H using the equation $\mu=B/H$, and we will obtain the curves in Figure 8. 2. 4. An instrument that will provide a plot of the B-H curve for a magnetic sample appears in Figure 8. 2. 7.

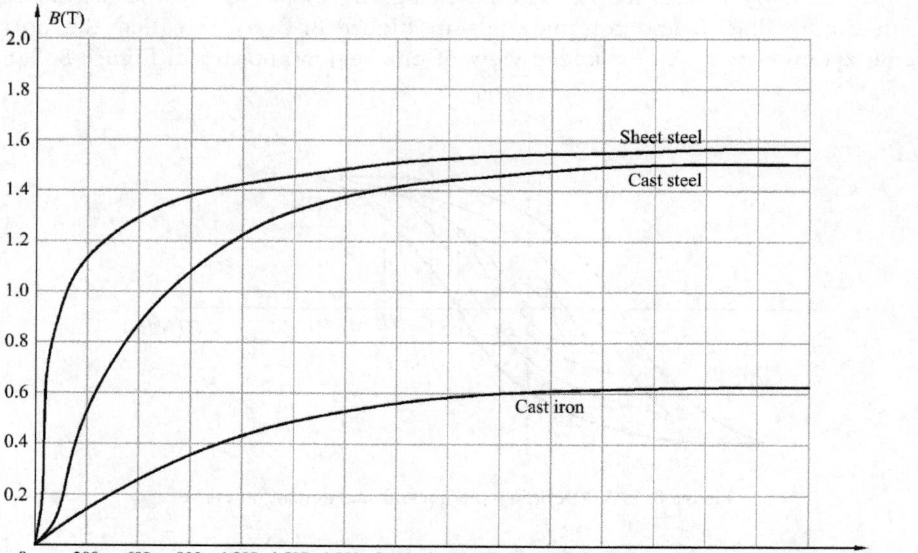

Figure 8. 2. 5 Normal magnetization curve for three ferromagnetic materials

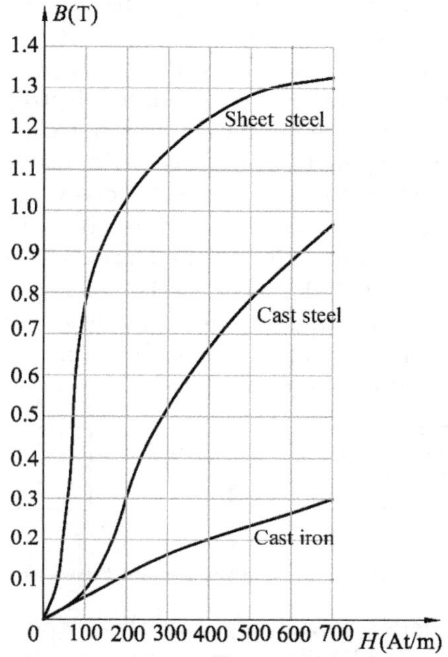

Figure 8. 2. 6 Expanding view of Figure 8. 2. 5 for the low magnetizing force region

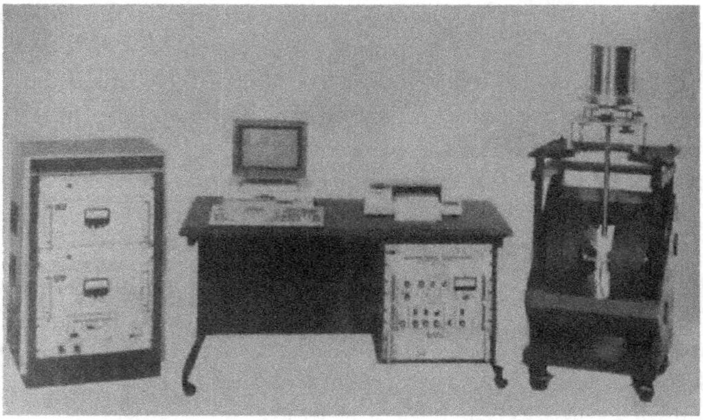

Figure 8. 2. 7 Model 9600 vibrating sample magnetometer. (Courtesy of LDJ Electronic, Inc)

It is interesting to note that the hysteresis curves in Figure 8. 2. 3 have a point **symmetry** about the origin; that is, the inverted pattern to the left of the vertical axis is the same as that appearing to the right of the vertical axis. In addition, you will find that a further application of the same magnetizing forces to the sample will result in the same plot. For a current i in $H=Ni/l$ that will move between positive and negative maximums at a fixed rate, the same B-H curve will result during each cycle. Such will be the case when we examine AC (sinusolidal) networks in the later chapters. The reversal of the field ϕ due to the changing current direction will result in a loss of energy that can best be described by first introducing the domain theory of magnetism.

Within each atom, the orbiting electrons are also spinning as they revolve around the nucleus. The atom, due to its spinning electrons, has a magnetic field associated with it. In nonmagnetic materials, the net magnetic field is effectively zero since the magnetic fields due to the atoms of the materials oppose each other. In magnetic materials such as iron and steel, however, the magnetic fields of groups of atoms numbering in the order of 10^{12} are aligned, forming very small bar magnets. This group of magnetically aligned atoms is called a **domain**. Each domain is a separate entity; that is, each domain is independent of the surrounding domains. For an unmagnetized sample of magnetic material, these domains appear in a random manner, such as shown in Figure 8.2.8(a). The net magnetic field in any direction is zero.

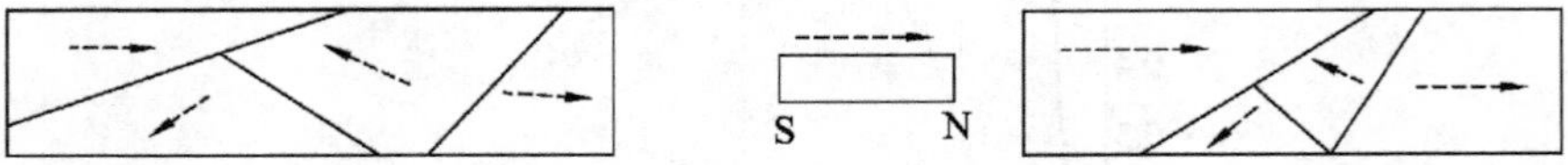

Figure 8.2.8 Demonstrating the domain theory of magnetism

When an external magnetizing force is applied, the domains that are nearly aligned with the applied field will grow at the expense of the less favorably oriented domains, such as shown in Figure 8.2.8(b). Eventually, if a sufficiently strong field is applied, all of the domains will have the orientation of the applied magnetizing force, and any further increase in external field will not increase the strength of the magnetic flux through the corea condition referred to as saturation. The elasticity of the above is evidenced by the fact that when the magnetizing force is removed, the alignment will be lost to some measure, and the flux density will drop to B_R. In other words, the removal of the magnetizing force will result in the return of a number of misaligned domains within the core. The continued alignment of a number of the domains, however, accounts for our ability to create permanent magnets.

At a point just before saturation, the opposing unaligned domains are reduced to small cylinders of various shapes referred to as bubbles. These bubbles can be moved within the magnetic sample through the application of a controlling magnetic field. These magnetic bubbles form the basis of the recently designed bubble memory system for computers.

8.3 Transformers

8.3.1 Mutual inductance

8.3.1.1 Mutual inductance

When two inductors (or coils) are in a close proximity to each other, the magnetic flux caused by current in one coil links with the other coil, thereby inducing voltage in the latter. This phenomenon is known as **mutual inductance.**

Mutual inductance results from a slight extension of this same argument. Firstly, let us consider a single inductor, a coil with N turns. When current i flows through the coil, a magnetic flux is produced around it (Figure 8.3.1). According to Faraday's law, the voltage v induced in the coil is proportional to the number of turns N and the time rate of change of the magnetic flux ϕ; that is,

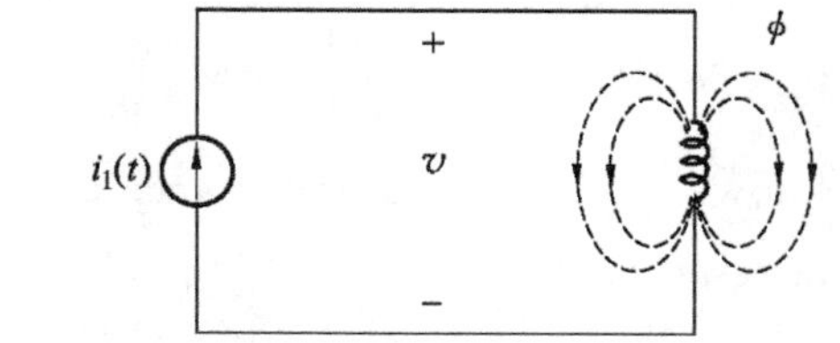

Figure 8. 3. 1 Magnetic flux produced by a single coil with N turns

$$v = N\frac{\mathrm{d}\phi}{\mathrm{d}t} \tag{8. 3. 1}$$

But the flux ϕ is produced by current i so that any change in ϕ is caused by a change in the current. Hence, Equation (8. 3. 1) can be written as

$$v = N\frac{\mathrm{d}\phi}{\mathrm{d}i}\frac{\mathrm{d}i}{\mathrm{d}t} = L\frac{\mathrm{d}i}{\mathrm{d}t} \tag{8. 3. 2}$$

which is the voltage-current relationship for the inductor. From Equation (8. 3. 2), the inductance L of the inductor is thus given by

$$L = N\frac{\mathrm{d}\phi}{\mathrm{d}t} \tag{8. 3. 3}$$

This inductance is commonly called **self-inductance**, because it relates the voltage induced in a coil by a time-varying current in the same coil.

8. 3. 1. 2 Magnetically coupled

Then consider two coils with self-inductances L_1 and L_2 that are in close proximity with each other (Figure 8. 3. 2). Coil 1 has N_1 turns, while coil 2 has N_2 turns. For the sake of simplicity, assume that the second inductor carries no current. The magnetic flux ϕ_1 emanating from coil 1 has two components: one component ϕ_{11} links only coil 1, and the other component ϕ_{12} links both coils. Hence,

$$\phi_1 = \phi_{11} + \phi_{12} \tag{8. 3. 4}$$

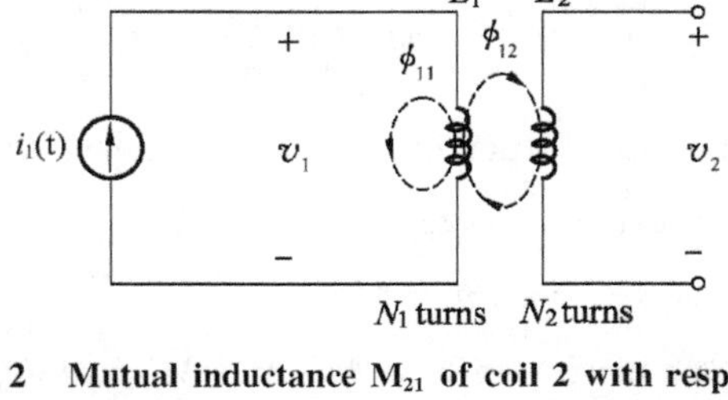

Figure 8. 3. 2 Mutual inductance M_{21} of coil 2 with respect to coil 1

Although the two coils are physically separated, they are said to be **magnetically coupled.** Since the entire flux ϕ_1 links coil 1 and the flux is caused by the current i_1 flowing in coil 1, the voltage induced in coil 1 is

$$v_1 = N_1\frac{\mathrm{d}\phi_1}{\mathrm{d}t} = N_1\frac{\mathrm{d}\phi_1}{\mathrm{d}i_1}\frac{\mathrm{d}i_1}{\mathrm{d}t} = L_1\frac{\mathrm{d}i_1}{\mathrm{d}t} \tag{8. 3. 5}$$

where $L_1 = N_1 \dfrac{d\phi_1}{di_1}$ is the self-inductance of coil 1. Similarly, only flux ϕ_{12} links coil 2, so the voltage induced in coil 2 is

$$v_2 = N_2 \frac{d\phi_{12}}{dt} = N_2 \frac{d\phi_{12}}{di_1}\frac{di_1}{dt} = M_{21}\frac{di_1}{dt} \tag{8.3.6}$$

where $M_{21} = N_2 \dfrac{d\phi_{12}}{di_1}$ is known as the mutual inductance of coil 2 with respect to coil 1. Subscript 21 indicates that the inductance M_{21} relates the voltage induced in coil 2 to the current in coil 1. Thus, the open-circuit mutual voltage (or induced voltage) across coil 2 is

$$v_2 = M_{21}\frac{di_1}{dt} \tag{8.3.7}$$

Suppose we now let current i_2 flow in coil 2, while coil 1 carries no current (Figure 8.3.3). The magnetic flux ϕ_2 emanating from coil 2 comprises flux ϕ_{22} that links only coil 2 and flux ϕ_{21} that links both coils. Hence,

$$\phi_2 = \phi_{22} + \phi_{21} \tag{8.3.8}$$

The entire flux ϕ_2 links coil 2, so the voltage induced in coil 2 is

$$v_2 = N_2 \frac{d\phi_2}{dt} = N_2 \frac{d\phi_2}{di_2}\frac{di_2}{dt} = L_2\frac{di_2}{dt} \tag{8.3.9}$$

where $L_2 = N_2 \dfrac{d\phi_2}{di_2}$ is the self-inductance of coil 2.

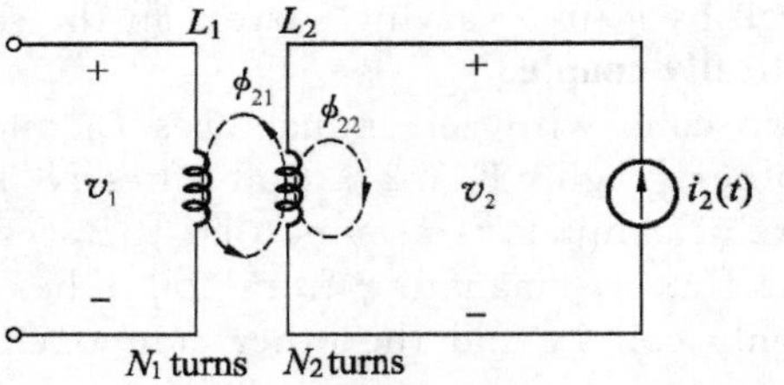

Figure 8.3.3 Mutual inductance M_{12} of coil 1 with respect to coil 2.

Since only flux links coil 1, the voltage induced in coil 1 is

$$v_1 = N_1 \frac{d\phi_{21}}{dt} = N_1 \frac{d\phi_{21}}{di_2}\frac{di_2}{dt} = M_{12}\frac{di_2}{dt} \tag{8.3.10}$$

where $M_{12} = N_1 \dfrac{d\phi_{21}}{di_2}$ is the mutual inductance of coil 1 with respect to coil 2. Thus, the open-circuit mutual voltage across coil 1 is

$$v_1 = M_{12}\frac{di_2}{dt} \tag{8.3.11}$$

We will see in the next section that M_{12} and M_{21} are equal; that is,

$$M_{12} = M_{21} = M \tag{8.3.12}$$

We refer to M as the mutual inductance between the two coils. Like self-inductance L, mutual inductance M is measured in henrys (H). Keep in mind that mutual coupling only exists when the inductors or coils are in close proximity, and the circuits are driven by time-varying sources.

8.3.1.3 Dot convention

Although mutual inductance M is always a positive quantity, the mutual voltage $M\mathrm{d}i/\mathrm{d}t$ may be negative or positive, just like the self-induced voltage $L\mathrm{d}i/\mathrm{d}t$. However, unlike the self-induced $L\mathrm{d}i/\mathrm{d}t$ whose polarity is determined by the reference direction of the current and the reference polarity of the voltage (according to the passive sign convention), the polarity of mutual voltage $M\mathrm{d}i/\mathrm{d}t$ is not easy to determine, because four terminals are involved. The choice of the correct polarity for $M\frac{\mathrm{d}i}{\mathrm{d}t}$ is made by examining the orientation or particular way in which both coils are physically wound and applying Lenz's law in conjunction with the right-hand rule. Since it is inconvenient to show the construction details of coils on a circuit schematic, we apply the **dot convention** in circuit analysis. By this convention, a dot is placed in the circuit at one end of each of the two magnetically coupled coils to indicate the direction of the magnetic flux if current enters that dotted terminal of the coil. This is illustrated in Figure 8.3.4. Given a circuit, the dots are already placed beside the coils so that we need not bother about how to place them. The dots are used along with the dot convention to determine the polarity of the mutual voltage. The dot convention is stated as follows:

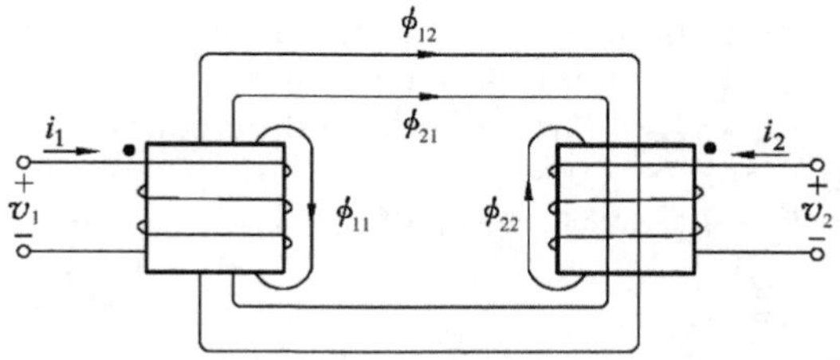

Figure 8.3.4 Illustration of the dot convention

A current entering the dotted terminal of one coil produces an open circuit voltage with a positive voltage reference at the dotted terminal of the second coil.

A current entering the undotted terminal of one coil provides a voltage that is positively sensed at the undotted terminal of the second coil.

For the coupled coils in Figure 8.3.5(a), the sign of the mutual voltage v_2 is determined by the reference polarity for v_2 and the direction of i_1. Since i_1 enters the dotted terminal of coil 1 and v_2 is positive at the dotted terminal of coil 2, the mutual voltage is $+M\mathrm{d}i/\mathrm{d}t$. For the coils in Figure 8.3.5 (b), the current i_1 enters the dotted terminal of coil 1 and v_2 is negative at the dotted terminal of coil 2. Hence, the mutual voltage is $-M\mathrm{d}i/\mathrm{d}t$. The same reasoning applies to the coils in Figure 8.3.5 (c) and 8.3.5 (d).

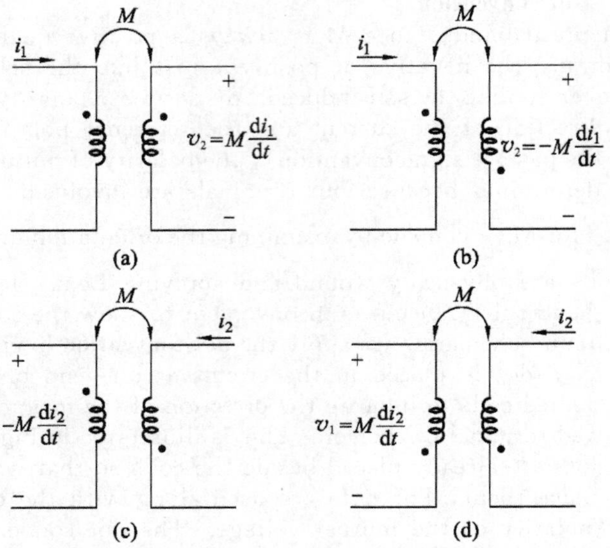

Figure 8.3.5 Examples illustrating how to apply the dot convention.

Figure 8.3.6 shows the dot convention for coupled coils in series. For the coils in Figure 8.3.6(a), the total inductance is

$$L=L_1+L_2+2M \quad \text{(Series-aiding connection)} \tag{8.3.13}$$

For the coils in Figure 8.3.6(b),

$$L=L_1+L_2-2M \quad \text{(Series-opposing connection)} \tag{8.3.14}$$

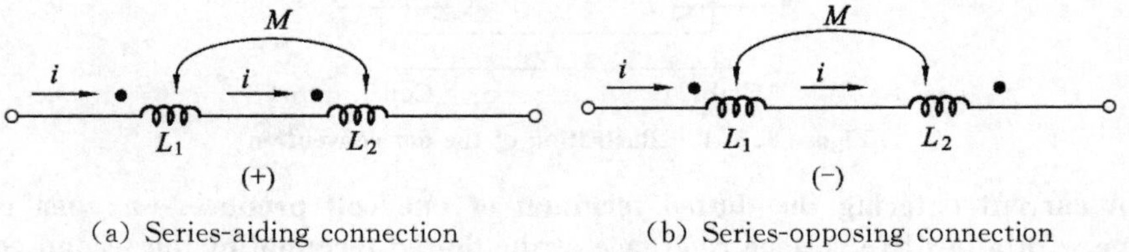
M
i
i
L_1
L_2
(+)
(a) Series-aiding connection
M
i
i
L_1
L_2
(−)
(b) Series-opposing connection

Figure 8.3.6 Dot convention for coils in series; the sign indicates the polarity of the mutual voltage

Now that we know how to determine the polarity of the mutual voltage, we are prepared to analyze circuits involving mutual inductance. As the first example, consider the circuit in Figure 8.3.7. Applying KVL to coil 1 gives

$$v_1=i_1R_1+L_1\frac{di_1}{dt}+M\frac{di_2}{dt} \tag{8.3.15a}$$

For coil 2, KVL gives

$$v_2=i_2R_2+L_2\frac{di_2}{dt}+M\frac{di_1}{dt} \tag{8.3.15b}$$

We can write Equation (8.3.15) in the frequency domain as

$$V_1=(R_1+\text{j}wL_1)I_1+\text{j}wMI_2 \tag{8.3.16a}$$

$$V_2=\text{j}wMI_1+(R_2+\text{j}wL_2)I_2 \tag{8.3.16b}$$

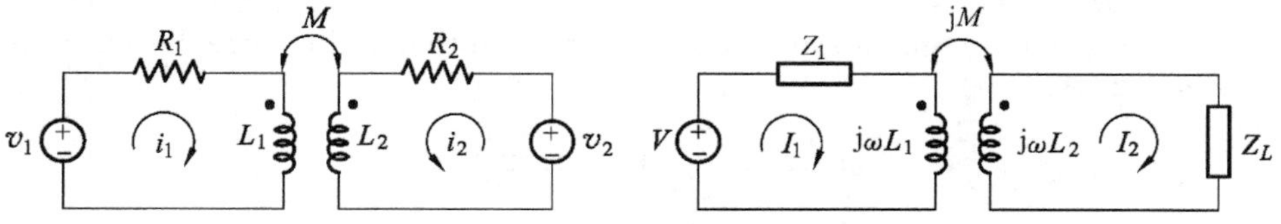

(a) Time-domain analysis (b) Frequency-domain analysis

Figure 8.3.7 Analysis of a circuit containing coupled coils

As another example, we consider the circuit in Figure 8.3.7(b). We analyze this in the frequency domain. Applying KVL to coil 1, we get

$$V=(Z_1+j\omega L_1)I_1-j\omega MI_2 \tag{8.3.17a}$$

For coil 2, KVL yields

$$0=-j\omega MI_1+(Z_L+j\omega L_2)I_2 \tag{8.3.17b}$$

Equations (8.3.16) and (8.3.17) are solved in the usual manner to determine the currents.

Example 8.3.1

Calculate the phasor currents I_1 and I_2 in the circuit of Figure for example 8.3.1.

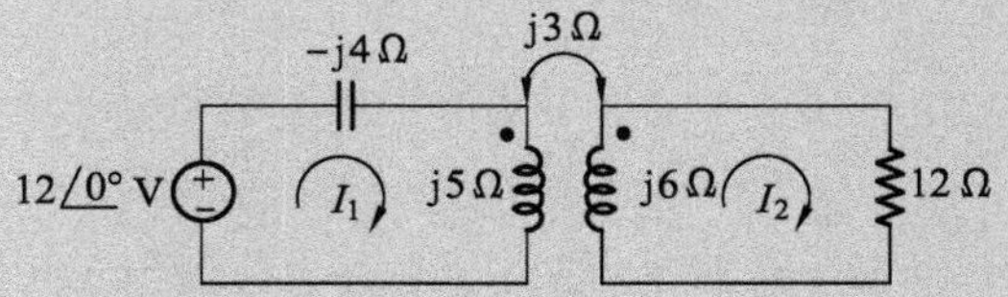

Figure for example **8.3.1**

Solution: For coil 1, KVL gives

$$-12+(-j4+j5)I_1-j3I_2=0$$

or

$$jI_1-j3I_2=12$$

For coil 2, KVL gives

$$-j3I_1+(12+j6)I_2=0$$

or

$$I_1=\frac{(12+j6)I_2}{j3}=(2-j4)I_2$$

Substituting I_1 in equation of I_2, we get $(j2+4-j3)I_2=(4-j)I_2=12$ or

$$I_2=\frac{12}{4-j}=2.91\angle 14.04^\circ \text{ A}$$

Thus,

$I_1=(2-j4)I_2=(4.472\angle -63.43^\circ)(2.91\angle 14.04^\circ)=13.01\angle -49.39^\circ$ A

8.3.1.4 Energy considerations in a coupled circuit

Let us now consider the energy stored in a pair of mutually coupled inductors. We will first justify our assumption that $M_{12}=M_{21}$, and we may then determine the maximum possible value of the mutual inductance between two given inductors.

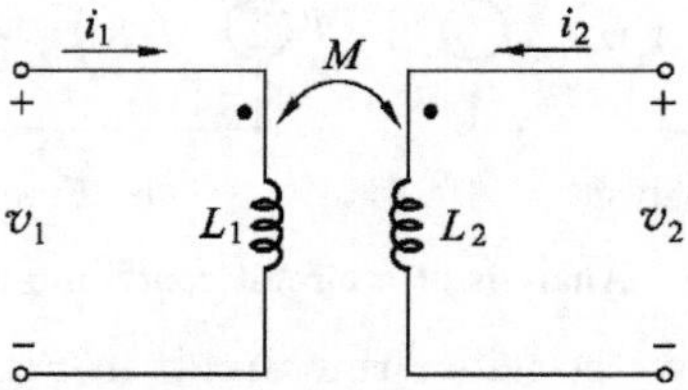

Figure 8. 3. 8 A pair of coupled coils with a mutual inductance of $M_{12}=M_{21}=M$

The pair of coupled coils shown in Figure 8. 3. 8 has currents, voltages, and polarity dots indicated. In order to show that $M_{12}=M_{21}$, we begin by letting all currents and voltages be zero, thus establishing zero initial energy storage in the network. Then, we open-circuit the right-hand terminal pair and increase i_1 from zero to some constant (DC) value I_1 at time $t=t_1$. The power entering the network from the left at any instant is

$$v_1 i_1 = L_1 \frac{di_1}{dt} i_1 \tag{8. 3. 18}$$

and the power entering from the right is

$$v_2 i_2 = 0 \tag{8. 3. 19}$$

since $i_2=0$.

Thus, the energy stored within the network when $i_1=I_1$ is

$$\int_0^{t_1} v_1 i_1 \, dt = \int_0^{I_1} L_1 i_1 \, di_1 = \frac{1}{2} L_1 I_1^2 \tag{8. 3. 20}$$

We now hold i_1 constant ($i_1=I_1$), and we let i_2 change from zero at $t=t_1$ to some constant value I_2 at $t=t_2$. The energy delivered from the right-hand source is thus

$$\int_{t_1}^{t_2} v_2 i_2 \, dt = \int_0^{I_2} L_2 i_2 \, di_2 = \frac{1}{2} L_2 I_2^2 \tag{8. 3. 21}$$

However, even though the value of i_1 remains constant, the left-hand source also delivers energy to the network during this time interval:

$$\int_{t_1}^{t_2} v_1 i_1 \, dt = \int_{t_1}^{t_2} M_{12} \frac{di_2}{dt} i_1 \, dt = M_{12} I_1 \int_0^{I_2} di_2 = M_{12} I_1 I_2 \tag{8. 3. 22}$$

The total energy stored in the network when both i_1 and i_2 have reached constant values is

$$W_{\text{total}} = \frac{1}{2} L_1 I_1^2 + \frac{1}{2} L_2 I_2^2 + M_{12} I_1 I_2 \tag{8. 3. 23}$$

Now, we may establish the same final currents in this network by allowing the currents to reach their final values in the reverse order, that is, first increasing i_2 from zero to I_2 and then holding i_2 constant while i_1 increases from zero to I_1. If the total energy stored is calculated for this experiment, the result is found to be

$$W_{\text{total}} = \frac{1}{2} L_1 I_1^2 + \frac{1}{2} L_2 I_2^2 + M_{21} I_1 I_2 \tag{8. 3. 24}$$

The only difference is the interchange of the mutual inductances M_{21} and

M_{12}. The initial and final conditions in the network are the same, however, and so the two values of the stored energy must be identical. Thus,

$$M_{12}=M_{21}=M \tag{8.3.25}$$

and

$$W=\frac{1}{2}L_1 I_1^2+\frac{1}{2}L_2 I_2^2+MI_1 I_2 \tag{8.3.26}$$

If one current enters a dot-marked terminal while the other leaves a dot-marked terminal, the sign of the mutual energy term is reversed:

$$W=\frac{1}{2}L_1 I_1^2+\frac{1}{2}L_2 I_2^2-MI_1 I_2 \tag{8.3.27}$$

Although Equations (8.3.26) and (8.3.27) were derived by treating the final values of the two currents as constants, these "constants" can have any value, and the energy expressions correctly represent the energy stored when the instantaneous values of i_1 and i_2 are I_1 and I_2, respectively. In other words, lowercase symbols might just as well be used:

$$w(t)=\frac{1}{2}L_1[i_1(t)]^2+\frac{1}{2}L_2[i_2(t)]^2\pm M[i_1(t)][i_2(t)] \tag{8.3.28}$$

The only assumption upon which Equation (8.3.28) is based is the logical establishment of a zero-energy reference level when both currents are zero.

Equation (8.3.28) may now be used to establish an upper limit for the value of M. Since $w(t)$ represents the energy stored within a passive network, it cannot be negative for any value of i_1, i_2, I_1, I_2, or M. Let us assume first that i_1 and i_2 are either both positive or both negative; their product is therefore positive. From Equation (8.3.28), the only case in which the energy could possibly be negative is

$$w=\frac{1}{2}L_1 i_1^2+\frac{1}{2}L_2 i_2^2-Mi_1 i_2 \tag{8.3.29}$$

which we may write, by completing the square, as

$$w=\frac{1}{2}(\sqrt{L_1}i_1-\sqrt{L_2}i_2)^2+\sqrt{L_1L_2}i_1 i_2-Mi_1 i_2 \tag{8.3.30}$$

Since in reality the energy cannot be negative, the right-hand side of this equation cannot be negative. The first term, however, may be as small as zero, so we have the restriction that the sum of the last two terms cannot be negative. Hence,

$$M\leqslant\sqrt{L_1L_2} \tag{8.3.31}$$

There is, therefore, an upper limit to the possible magnitude of the mutual inductance; it can be no larger than the geometric mean of the inductances of the two coils between which the mutual inductance exists. Although we have derived this inequality on the assumption that i_1 and i_2 carried the same algebraic sign, a similar development is possible if the signs are opposite; it is necessary only to select the positive sign in Equation (8.3.28).

We might also have demonstrated the truth of Equation (8.3.31) from a physical consideration of the magnetic coupling; if we think of i_2 as being zero and the current i_1 as establishing the magnetic flux linking both L_1 and L_2, it is apparent that the flux within L_2 cannot be greater than the flux within L_1, which represents the total flux. Qualitatively, then, there is an upper limit to the magnitude of the mutual inductance possible between two given inductors.

The degree to which M approaches its maximum value is described by the **coupling coefficient**, defined as

$$k=\frac{M}{\sqrt{L_1 L_2}} \qquad (8.3.32)$$

since $M \leqslant \sqrt{L_1 L_2}$,

$$0 \leqslant k \leqslant 1$$

The **coupling coefficient** k is a measure of the magnetic coupling between two coils. We expect k to depend on the closeness of the two coils, their core, their orientation, and their windings. If the entire flux produced by one coil links another coil, then $k=1$ and we have 100% coupling, or the coils are said to be **perfectly coupled.** For $k<0.5$, coils are said to be **loosely coupled**; and for $k>0.5$, they are said to be **tightly coupled.** Figure 8.3.9 shows loosely coupled windings and tightly coupled windings. The air-core transformers used in radio frequency circuits are loosely coupled, whereas iron-core transformers used in power systems are tightly coupled. The linear transformers discussed are mostly air-core; the ideal transformers discussed are principally iron-core.

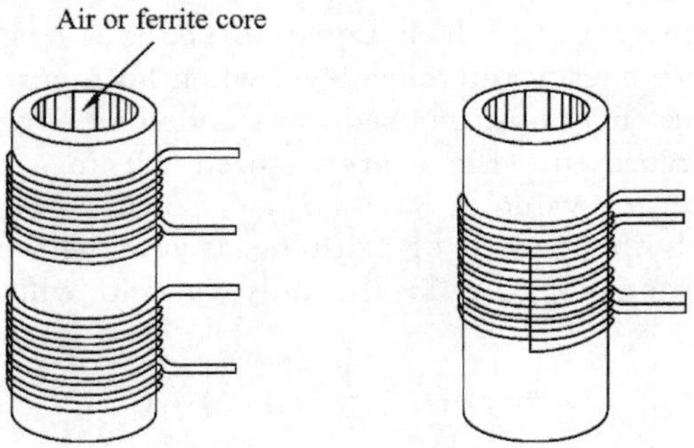

(a) Loosely coupled (b) Tightly coupled

Figure 8.3.9 Cutaway view demonstrates both windings

8.3.2 Linear transformer

Here we introduce the transformer as a new circuit element. The **transformer** is a device that couples two AC circuits magnetically rather than through any direct conductive connection and permits a "transformation" of the voltage and current between one circuit and the other (for example, by matching a high-voltage, low-current AC output to a circuit requiring a low-voltage, high-current source). Transformers play a major role in electric power engineering and are a necessary part of the electric power distribution network.

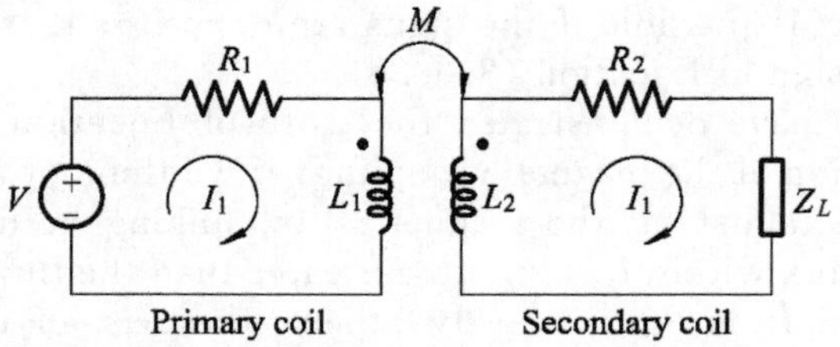

Figure 8.3.10 A linear transformer

In Figure 8. 3. 10 a transformer is shown with two mesh currents identified. The first mesh, usually containing the source, is called the primary, while the second mesh, usually containing the load, is known as the secondary. The inductors labeled L_1 and L_2 are also referred to as **the primary** coil and **secondary** coil of the transformer, respectively. We will assume that the transformer is linear. This implies that no magnetic material (which may cause a nonlinear flux-versus-current relationship) is employed. Without such material, however, it is difficult to achieve a coupling coefficient greater than a few tenths. The two resistors serve to account for the resistance of the wire out of which the primary and secondary coils are wound, and any other loss.

Consider the input impedance offered at the terminals of the primary circuit. The two mesh equations are

$$V=(R_1+\mathrm{j}wL_1)I_1-\mathrm{j}wMI_2 \qquad (8.3.33a)$$

$$0=-\mathrm{j}wMI_1+(R_2+\mathrm{j}wL_2+Z_L)I_2 \qquad (8.3.33b)$$

In Equation (8. 3. 33b), we express I_2 in terms of I_1 and substitute it into Equation (8. 3. 33a). We get the input impedance as

$$Z_{\text{in}}=\frac{V}{I}=\mathrm{R}_1+\mathrm{j}wL_1+\frac{w^2M^2}{R_2+\mathrm{j}wL_2+Z_L} \qquad (8.3.34)$$

Notice that the input impedance comprises two terms. The first term, $(R_1+\mathrm{j}wL_1)$, is the primary impedance. The second term is due to the coupling between the primary and secondary windings. It is as though this impedance is reflected to the primary. Thus, it is known as the **reflected impedance** Z_{R} and

$$Z_{\mathrm{R}}=\frac{w^2M^2}{R_2+\mathrm{j}wL_2+Z_L} \qquad (8.3.35)$$

Analyzing magnetically coupled circuits is not as easy as circuits in previous chapters. For this reason, it is sometimes convenient to replace a magnetically coupled circuit by an equivalent circuit with no magnetic coupling. We want to replace the linear transformer in Figure 8. 3. 10 by an equivalent T or Π circuit, a circuit that would have no mutual inductance.

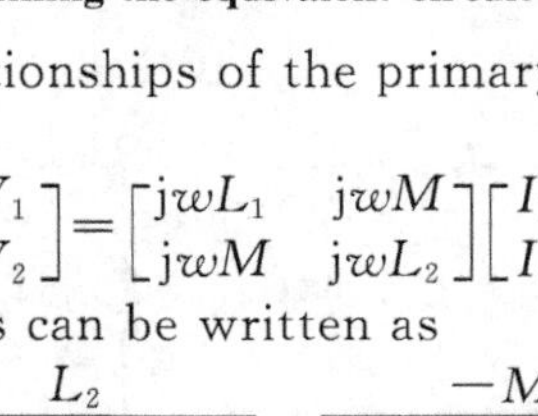

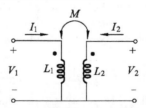

Figure 8. 3. 11 Determining the equivalent circuit of a linear transformer

The voltage-current relationships of the primary and secondary coils give the matrix equation.

$$\begin{bmatrix} V_1 \\ V_2 \end{bmatrix}=\begin{bmatrix} \mathrm{j}wL_1 & \mathrm{j}wM \\ \mathrm{j}wM & \mathrm{j}wL_2 \end{bmatrix}\begin{bmatrix} I_1 \\ I_2 \end{bmatrix} \qquad (8.3.36)$$

By matrix inversion, this can be written as

$$\begin{bmatrix} I_1 \\ I_2 \end{bmatrix}=\begin{bmatrix} \dfrac{L_2}{\mathrm{j}w(L_1L_2-M^2)} & \dfrac{-M}{\mathrm{j}w(L_1L_2-M^2)} \\ \dfrac{-M}{\mathrm{j}w(L_1L_2-M^2)} & \dfrac{L_1}{\mathrm{j}w(L_1L_2-M^2)} \end{bmatrix}\begin{bmatrix} V_1 \\ V_2 \end{bmatrix} \qquad (8.3.37)$$

Our goal is to match Equations (8. 3. 36) and (8. 3. 37) with the corresponding equations for the T and networks.

For the T (or Y) network of Figure 8. 3. 12, mesh analysis provides the terminal equations as

$$\begin{bmatrix} V_1 \\ V_2 \end{bmatrix}=\begin{bmatrix} jw(L_a+L_c) & jwL_c \\ jwL_c & jw(L_b+L_c) \end{bmatrix}\begin{bmatrix} I_1 \\ I_2 \end{bmatrix} \tag{8. 3. 38}$$

If the circuits in Figures 8. 3. 11 and 8. 3. 12 are equivalents, Equations (8. 3. 36) and (8. 3. 38) must be identical. Equating terms in the impedance matrices of Equations (8. 3. 36) and (8. 3. 38) leads to

$$L_a=L_1-M,\ L_b=L_2-M,\ L_c=M \tag{8. 3. 39}$$

For the Π (or Δ) network in Figures 8. 3. 13, nodal analysis gives the terminal equations as

$$\begin{bmatrix} I_1 \\ I_2 \end{bmatrix}=\begin{bmatrix} \frac{1}{jwL_A}+\frac{1}{jwL_C} & -\frac{1}{jwL_C} \\ -\frac{1}{jwL_C} & \frac{1}{jwL_B}+\frac{1}{jwL_C} \end{bmatrix}\begin{bmatrix} V_1 \\ V_2 \end{bmatrix} \tag{8. 3. 40}$$

Equating terms in admittance matrices of Equations (8. 3. 37) and (8. 3. 40), we obtain

$$L_A=\frac{L_1L_2-M}{L_2-M},\ L_B=\frac{L_1L_2-M}{L_1-M},\ L_C=\frac{L_1L_2-M^2}{M} \tag{8. 3. 41}$$

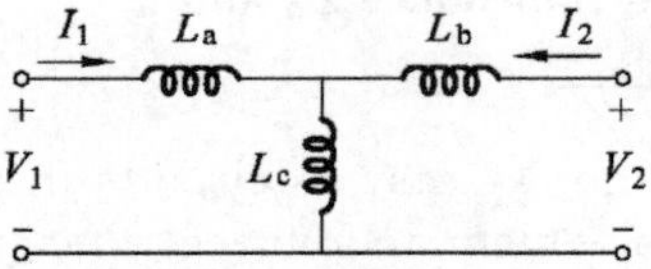

Figure 8. 3. 12 An equivalent T circuit

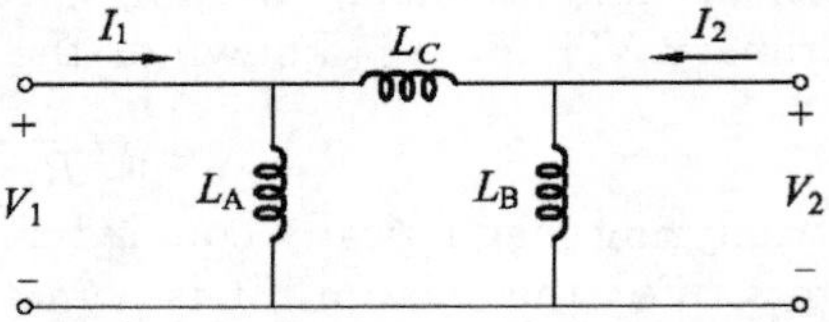

Figure 8. 3. 13 An equivalent Π circuit

Note that in Figures 8. 3. 12 and 8. 3. 13, the inductors are not magnetically coupled. Also note that changing the locations of the dots in Figure 8. 3. 13 can cause M to become $-M$. As Example 8. 3. 2 illustrates, a negative value of M is physically unrealizable but the equivalent model is still mathematically valid.

Example 8.3.2

Solve for I_1, I_2 and V_0 in Figure A for example 8.3.2 using the T-equivalent circuit for the linear transformer.

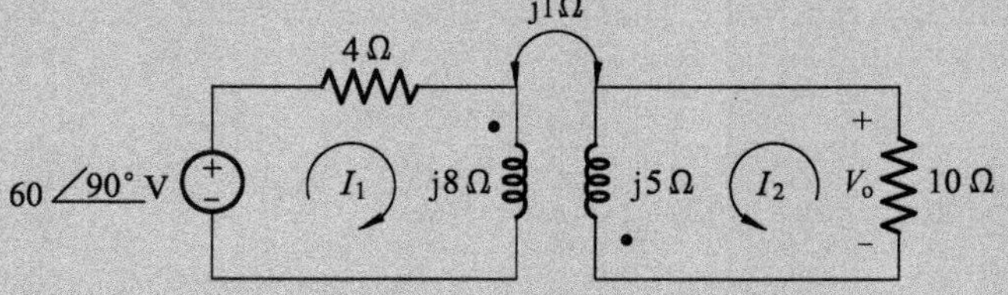

Figure A for example 8. 3. 2

Solution: We need to replace the magnetically coupled coils with the T-equivalent circuit. The relevant portion of the circuit in Figure A for example 8. 3. 2 is shown in the following Figure B for example 8. 3. 2(a). Comparing the Figure B for example 8. 3. 2 (a) with the figure above shows that there are two differences. First, due to the current reference directions and voltage polarities, we need to replace M by $-M$ to make Figure B for example 8. 3. 2(a) conform with Figure 8. 3. 11. Second, the circuit in the figure above is in the time-domain, whereas the circuit in Figure B for example 8. 3. 2(a) is in the frequency-domain. The difference is the factor jw, that is, L in the figure above has been replaced with jwL and M with jwM. Since w is not specified, we can assume $w=1$ rad/s or any other value; it really does not matter. With these two differences in mind, $L_a=L_1-(-M)=8+1=9$ H, $L_b=L_2-(-M)=5+1=6$ H, $L_c=-M=-1$ H

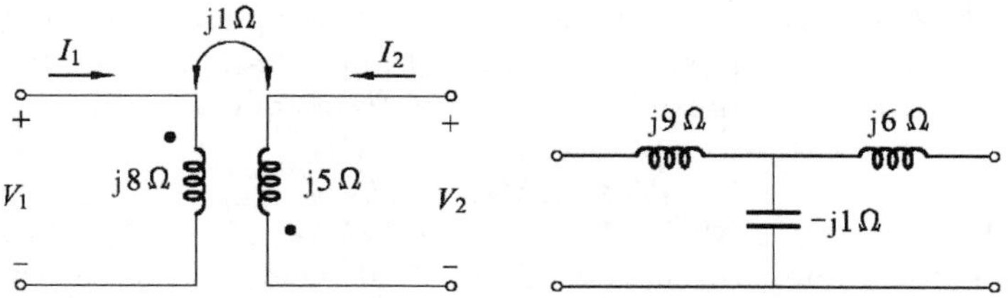

(a) Circuit for coupled coils of Figure (b) T-equivalent circuit

Figure B for example 8. 3. 2

Thus, the T-equivalent circuit for the coupled coils is as shown in Figure B for example 8. 3. 2(b).

Inserting the T-equivalent circuit in Figure B for example 8. 3. 2(b) to replace the two coils in figure, A for example 8. 3. 2 gives the equivalent circuit in the following figure, which can be solved using nodal or mesh analysis. Applying mesh analysis, we obtain

$$\mathrm{j}6=I_1(4+\mathrm{j}9-\mathrm{j}1)+I_2(-\mathrm{j}1)$$

and

$$0=I_1(-\mathrm{j}1)+I_2(10+\mathrm{j}6-\mathrm{j}1)$$

Then,

$$I_1=\frac{(10+\mathrm{j}5)}{\mathrm{j}}I_2=(5-\mathrm{j}10)I_2$$

Substituting equation of I_1 into the first equation gives

$$\mathrm{j}6=(4+\mathrm{j}8)(5-\mathrm{j}10)I_2-\mathrm{j}I_2=(100-\mathrm{j})I_2\approx 100I_2$$

Since 100 is very large compared with 1, the imaginary part of $(100-\mathrm{j})$ can be ignored so that $100-\mathrm{j}\approx 100$. Hence,

$$I_2=\frac{\mathrm{j}6}{100}=\mathrm{j}0.06=0.06\angle 90^\circ \text{ A}$$

$$I_1=(5-\mathrm{j}10)\mathrm{j}0.06=(0.6+\mathrm{j}0.3) \text{ A}$$

and

$$V_0=-10I_2=-\mathrm{j}0.6=0.6\angle -90^\circ \text{ V}$$

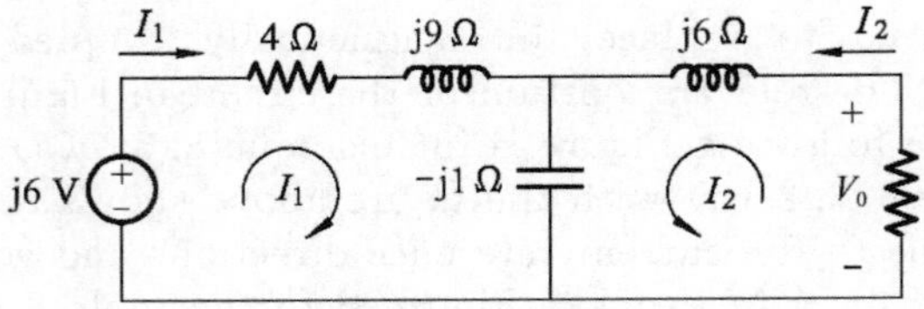

Figure C for example 8.3.2

8.3.3 The ideal transformer

An ideal transformer is one with perfect coupling ($k=1$). It consists of two (or more) coils with a large number of turns wound on a common core of high permeability. Because of this high permeability of the core, the flux links all the turns of both coils, thereby resulting in a perfect coupling.

To see how an ideal transformer is the limiting case of two coupled inductors where the inductances approach infinity and the coupling is perfect, let us reexamine the circuit in Figure 8.3.8. In the frequency domain,

$$V_1 = \mathrm{j}\omega L_1 I_1 + \mathrm{j}\omega M I_2 \tag{8.3.42a}$$

$$V_2 = \mathrm{j}\omega M I_1 + \mathrm{j}\omega L_2 I_2 \tag{8.3.42b}$$

From Equation (8.3.42a), $I_1 = (V_1 - \mathrm{j}\omega M I_2)/\mathrm{j}\omega L_1$ (we could have also use this equation to develop the current ratios instead of using the conservation of power which we will do shortly). Substituting this in Equation (8.3.42b) gives

$$V_2 = \mathrm{j}\omega L_2 I_2 + \frac{M V_1}{L_1} - \frac{\mathrm{j}\omega M^2 I_2}{L_1} \tag{8.3.43}$$

But $M = \sqrt{L_1 L_2}$ for perfect coupling ($k=1$). Hence,

$$V_2 = \mathrm{j}\omega L_2 I_2 + \frac{\sqrt{L_1 L_2} V_1}{L_1} - \frac{\mathrm{j}\omega L_1 L_2 I_2}{L_1} = \sqrt{\frac{L_2}{L_1}} V_1 = n V_1 \tag{8.3.44}$$

where $n = \sqrt{L_2 / L_1}$ and is called the turns ratio. As L_1, L_2, $M \to \infty$ such that n remains the same, the coupled coils become an ideal transformer.

A transformer is said to be ideal if it has the following properties:

(1) Coils have very large reactances(L_1, L_2, $M \to \infty$);

(2) Coupling coefficient is equal to unity($k=1$);

(3) Primary and secondary coils are lossless ($R_1 = R_2 = 0$).

An **ideal transformer** is a unity-coupled, lossless transformer in which the primary and secondary coils have infinite self-inductances.

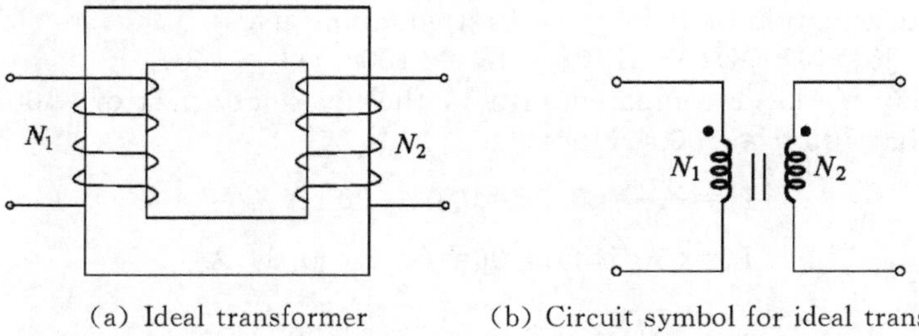

(a) Ideal transformer　　(b) Circuit symbol for ideal transformers

Figure 8.3.14　Ideal transformer

Figure 8.3.14(a) shows a typical ideal transformer; the circuit symbol is in Figure 8.3.14(b). The vertical lines between the coils indicate an iron core as distinct from the air core used in linear transformers. The primary winding has N_1 turns; the secondary winding has N_2 turns.

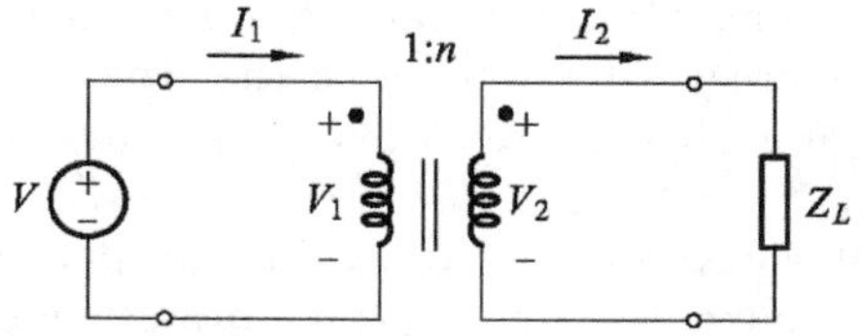

Figure 8.3.15 Relating primary and secondary quantities in an ideal transformer

When a sinusoidal voltage is applied to the primary winding as shown in Figure 8.3.15, the same magnetic flux ϕ goes through both windings. According to Faraday's law, the voltage across the primary winding is

$$v_1 = N_1 \frac{d\phi}{dt} \tag{8.3.45a}$$

while that across the secondary winding is

$$v_2 = N_2 \frac{d\phi}{dt} \tag{8.3.45b}$$

Dividing Equations (8.3.45b) by Equation (8.3.45a), we get

$$\frac{v_2}{v_1} = \frac{N_2}{N_1} = n \tag{8.3.46}$$

where n is, again, the turns ratio or transformation ratio. We can use the phasor voltages V_1 and V_2 rather than the instantaneous values v_1 and v_2. Thus, Equation (8.3.46) may be written as

$$\frac{V_2}{V_1} = \frac{N_2}{N_1} = n \tag{8.3.47}$$

For the reason of power conservation, the energy supplied to the primary coil must equal the energy absorbed by the secondary coil, since there are no losses in an ideal transformer. This implies that

$$v_1 i_1 = v_2 i_2 \tag{8.3.48}$$

In phasor form, Equation (8.3.48) in conjunction with Equation (8.3.47) becomes

$$\frac{I_1}{I_2} = \frac{V_1}{V_2} = n \tag{8.3.49}$$

showing that the primary and secondary currents are related to the turns ratio in the inverse manner as the voltages. Thus,

$$\frac{I_2}{I_1} = \frac{N_1}{N_2} = \frac{1}{n} \tag{8.3.50}$$

When $n = 1$, we generally call the transformer an **isolation transformer.** If $n > 1$, we have a **step-up transformer**, as the voltage is increased from primary to secondary ($V_2 > V_1$). On the other hand, if $n < 1$, the transformer is a **step-down transformer**, since the voltage is decreased from primary to secondary ($V_2 < V_1$).

A **step-down transformer** is one whose secondary voltage is less than its primary voltage.

A **step-up transformer** is one whose secondary voltage is greater than its primary voltage.

The ratings of transformers are usually specified as V_1/V_2. A transformer with rating 2 400/120 V should have 2 400 V in the primary and 120 V in the secondary (i. e. a step-down transformer). Keep in mind that the voltage ratings are in rms. Power companies often generate at some convenient voltage and use a step-up transformer to increase the voltage so that the power can be transmitted at very high voltage and low current over transmission lines, resulting in significant cost savings.

It is important that we know how to get the proper polarity of the voltages and the direction of the currents for the transformer in Figure 8. 3. 15. If the polarity of V_1 or V_2 or the direction of I_1 or I_2 is changed, n in Equations (8. 3. 46) to (8. 3. 50) may need to be replaced by $-n$. The two simple rules to follow are:

(1) If V_1 and V_2 are **both positive** or **both negative** at the dotted terminals, use $+n$ in Equation 8. 3. 47. Otherwise, use $-n$.

(2) If I_1 and I_2 **both enter** into or **both leave** the dotted terminals, use $-n$ in Equation (8. 3. 50). Otherwise, use $+n$.

The rules are demonstrated with the four circuits in Figure 8. 3. 16.

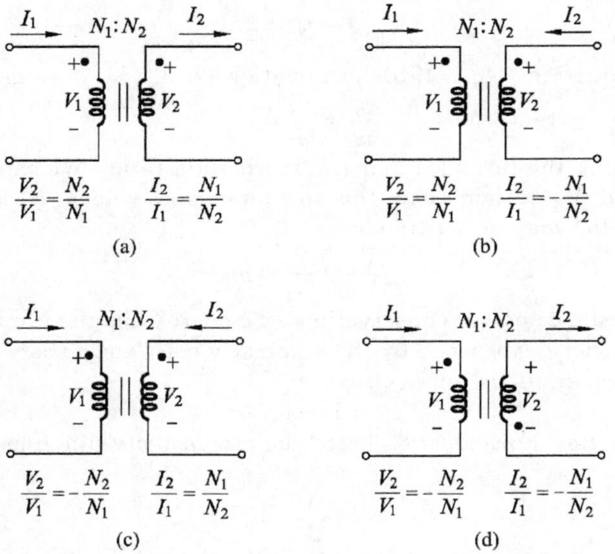

Figure 8. 3. 16 Typical circuits illustrating proper voltage polarities and current directions in an ideal transformer

Using Equations (8. 3. 47) and (8. 3. 50), we can always express V_1 in terms of V_2 and I_1 in terms of I_2, or vice versa:

$$V_1=\frac{V_2}{n},\ V_2=nV_1 \tag{8. 3. 51}$$

$$I_1=nI_2,\ I_2=\frac{1}{n}I_1 \tag{8. 3. 52}$$

The complex power in the primary winding is

$$S_1=V_1I_1^*=\frac{V_2}{n}(nI_2)^*=V_2I_2^*=S_2 \tag{8. 3. 53}$$

showing that the complex power supplied to the primary is delivered to the secondary without loss. The transformer absorbs no power. Of course, we should expect this, since the ideal transformer is lossless. The input impedance as seen by the source in Figure 8. 3. 15 is found from Equations (8. 3. 51) and (8. 3. 52) as

$$Z_{in}=\frac{V_1}{I_1}=\frac{1}{n^2}\frac{V_2}{I_2} \tag{8. 3. 54}$$

It is evident from Figure 8. 3. 15 that $V_2/I_2=Z_L$, so that

$$Z_{in}=\frac{Z_L}{n^2} \tag{8. 3. 55}$$

The input impedance is also called the **reflected impedance**, since it appears as if the load impedance is reflected to the primary side. This ability of the transformer to transform a given impedance into another impedance provides us with a means of **impedance matching** to ensure maximum power transfer.

In analyzing a circuit containing an ideal transformer, it is common practice to eliminate the transformer by reflecting impedances and sources from one side of the transformer to the other.

In the circuit of Figure 8. 3. 17, suppose we want to reflect the secondary side of the circuit to the primary side. We find the equivalent of the circuit to the right of the terminals *a*-*b*, and obtain V_{Th} as the open-circuit voltage at terminals *a*-*b*, as shown in Figure 8. 3. 18(a).

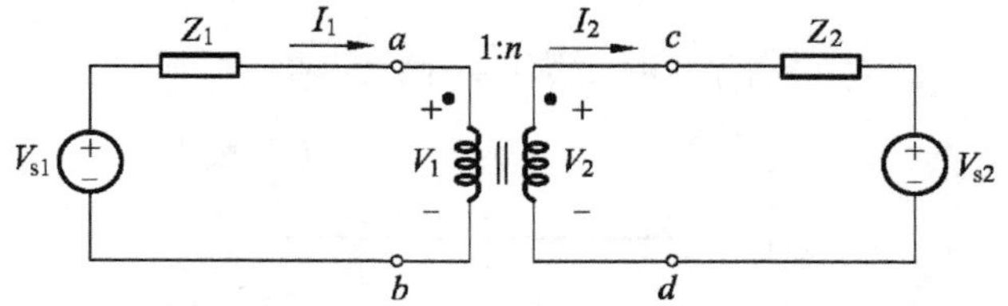

Figure 8. 3. 17 Ideal transformer circuit whose equivalent circuits are to be found

Since terminals *a*-*b* are open, $I_1=I_2=0$ so that $V_2=V_{s2}$. Hence, from Equation (8. 3. 51),

$$V_{Th}=V_1=\frac{V_2}{n}=\frac{V_{s2}}{n} \tag{8. 3. 56}$$

To get Z_{Th}, we remove the voltage source in the secondary winding and insert a unit source at terminals *a*-*b* as in Figure 8. 3. 18(b). From Equations (8. 3. 51) and (8. 3. 52), $I_1=nI_2$ and $V_1=V_2/n$, so that

$$Z_{Th}=\frac{V_1}{I_1}=\frac{V_2/n}{nI_2}=\frac{Z_2}{n^2},\ V_2=Z_2I_2 \tag{8. 3. 57}$$

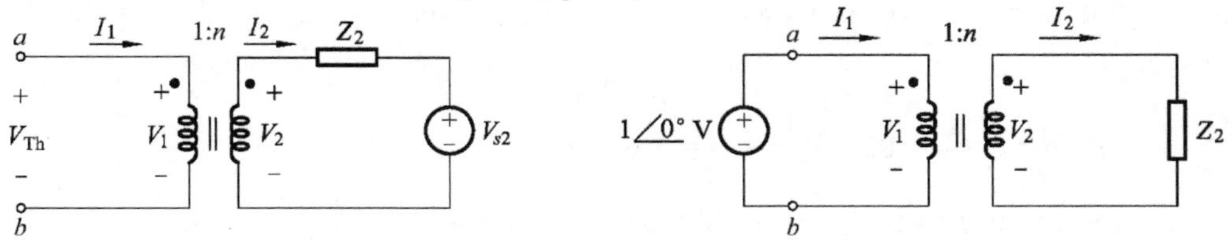

(a) Obtaining V_{Th} for the circuit in Figure 8. 3. 17 (b) Obtaining Z_{Th} for the circuit in Figure 8. 3. 17

Figure 8. 3. 18 Obtaining V_{Th} and Z_{Th} for the circuit in Figure 8. 3. 17

which is what we should have expected from Equation (8. 3. 55). Once we have and we add the equivalent to the part of the circuit in Figure 8. 3. 17 to the left of terminals *a-b*. Figure 8. 3. 19 shows the result.

The general rule of eliminating the transformer and reflecting the secondary circuit to the primary side is: divide the secondary impedance by n^2, divide the secondary voltage by n, and multiply the secondary current by n.

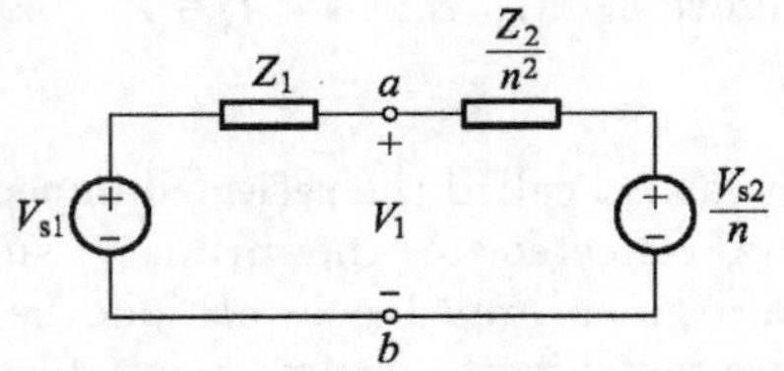

Figure 8. 3. 19 Equivalent circuit for Figure 8. 3. 17 obtained by reflecting the secondary circuit to the primary side.

We can also reflect the primary side of the circuit in Figure 8. 3. 17 to the secondary side. Figure 8. 3. 20 shows the equivalent circuit.

The rule of eliminating the transformer and reflecting the primary circuit to the secondary side is: multiply the primary impedance by n^2, multiply the primary voltage by n, and divide the primary current by n.

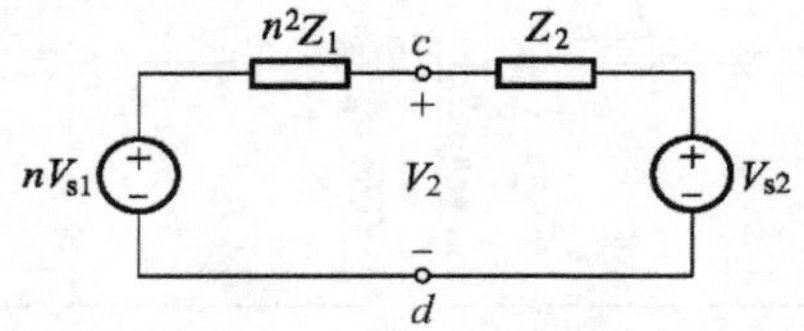

Figure 8. 3. 20 Equivalent circuit for Figure 8. 3. 17 obtained by reflecting the primary circuit to the secondary side

Example 8.3.3

An ideal transformer is rated at 2 400/120 V, 9.6 kVA, and has 50 turns on the secondary side. Calculate: (a) the turns ratio; (b) the number of turns on the primary side; (c) the current ratings for the primary and secondary windings.

Solution: (a) This is a step-down transformer, since $V_1=2\ 400\ \text{V}>V_2=120\ \text{V}$

$$n=\frac{V_2}{V_1}=\frac{120}{2\ 400}=0.05$$

(b) $n=\dfrac{N_2}{N_1}\Rightarrow 0.05=\dfrac{50}{N_1}\Rightarrow N_1=1\ 000$ turns

(c) $S=V_1I_1=V_2I_2=9.6$ kVA. Hence, $I_1=\dfrac{9\ 600}{V_1}=\dfrac{9\ 600}{2\ 400}=4\ \text{A}$

$$I_2=\frac{9\ 600}{V_2}=\frac{9\ 600}{120}=80\ \text{A or } I_2=\frac{I_1}{n}=\frac{4}{0.05}=80\ \text{A}$$

Summary

The objective of this chapter is to introduce magnetic circuit concepts and simple transformers calculation to the students.

1. Identify the basic concept of magnetic field and the law of magnetic circuit and it serves as the foundation to the transformers.

2. Present the magnetic materials, and make a deep analysis of the hysteresis curve.

3. Through the introduction of mutual inductance and magnetic coupling, the concept of transformer is proposed. The linear transformers and the ideal transformers are discussed in this chapter.

Problems

8.1 Determine the voltage V_0 in the circuit of Figure for problem 8.1.

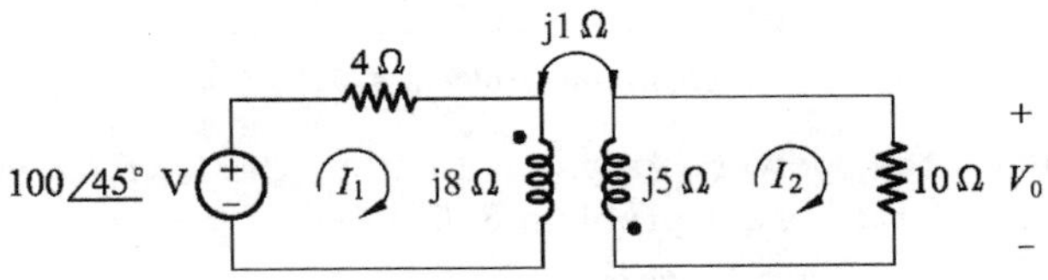

Figure for problem 8.1

8.2 Calculate the mesh currents in the circuit of Figure for problem 8.2.

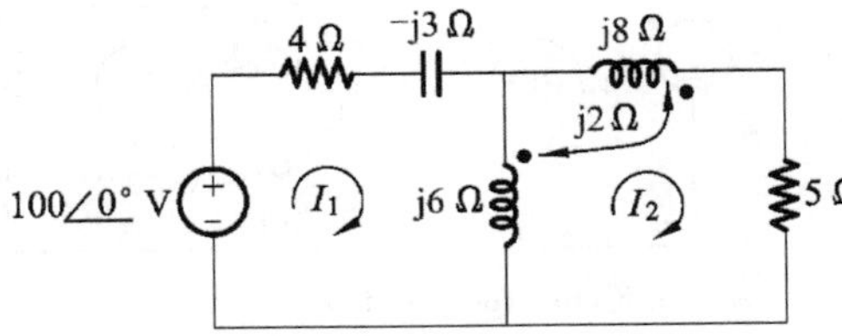

Figure for problem 8.2

8.3 Determine the phasor currents I_1 and I_2 in the circuit of Figure for problem 8.3.

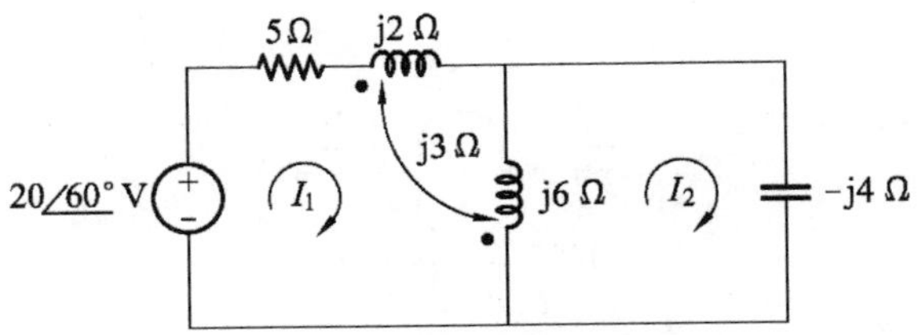

Figure for problem 8.3

8.4 In the circuit in Figure for problem 8.4, determine the coupling coefficient and the energy stored in the coupled inductors at $t=1.5$ s.

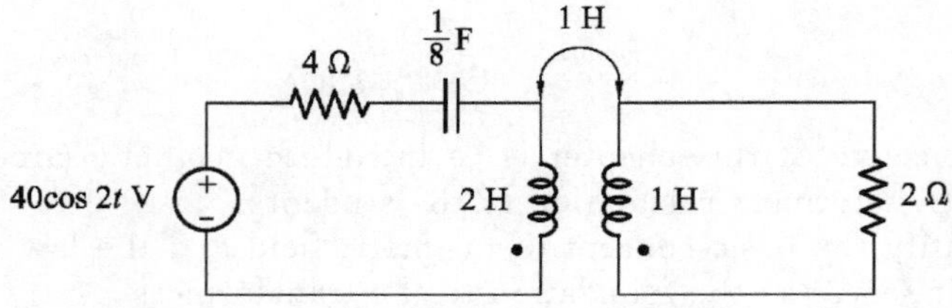

Figure for problem 8. 4

8. 5 Find V_0 in the circuit of Figure for problem 8. 5.

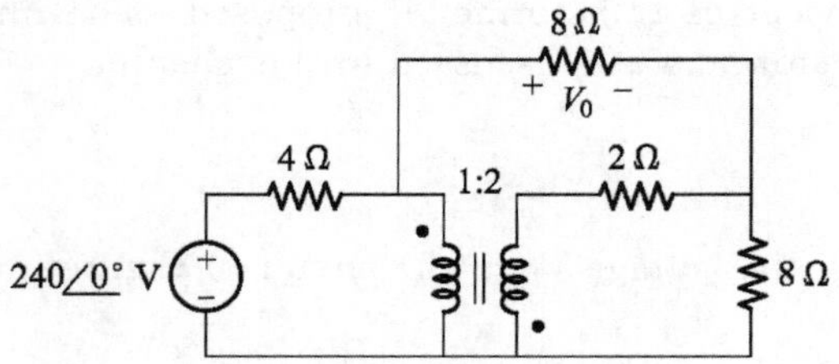

Figure for problem 8. 5

8. 6 Calculate the power supplied to the 10 Ω resister in the ideal transformer circuit of Figure for problem 8. 6.

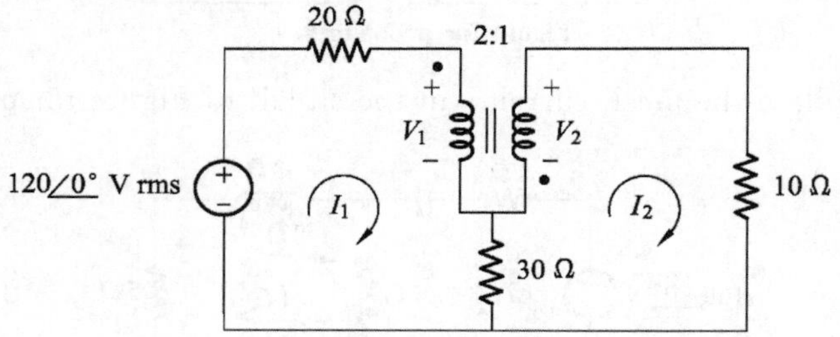

Figure for problem 8. 6

8. 7 The primary current to an ideal transformer rated at 3 300/110 V is 5 A. Calculate:

(a) the turns ratio; (b) the KVA rating; (c) the secondary current.

8. 8 Determine the T-equivalent circuit of the linear transformer in Figure for problem 8. 8.

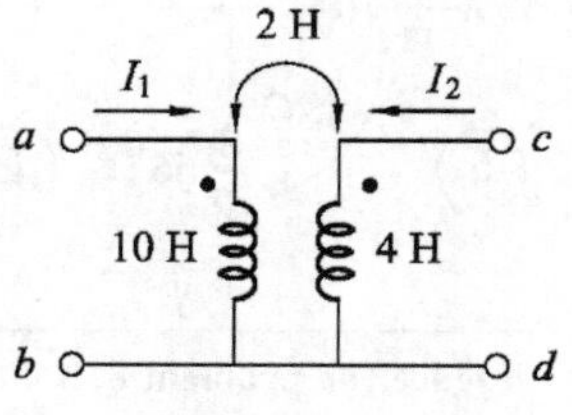

A linear transformer

Figure for problem 8. 8

CHAPTER 9
AC MACHINES

Introduction

Electric machines are broadly classified into DC and AC machines. The former use DC excitation for both the field and armature circuits, while the latter may be further subdivided into two classes: synchronous machines, and induction motors.

The AC induction motor is the workhorse of many industrial applications, and synchronous generators are used almost exclusively for the generation of electric power worldwide. Thus, it is appropriate to devote a significant portion of this chapter to the study of AC machines, and of induction motors in particular.

The principles developed in this chapter can be applied to rotating electric machines, to explain how mechanical energy can be converted to electrical energy, and vice versa. The former function is performed by electric generators, while the latter is provided by electric motors.

9.1 Basic classification of electric machines

An immediate distinction can be made between different types of windings characterized by the nature of the current they carry. If the current serves the sole purpose of providing a magnetic field and is independent of the load, it is called a **magnetizing or exciting current**, and the winding is termed a **field winding.** Field currents are nearly always DC and are of relatively low power, since their only purpose is to magnetize the core. On the other hand, if the winding carries only the load current, it is called an **armature.** In DC and AC synchronous machines, separate windings exist to carry field and armature currents. In the induction motor, the magnetizing and load currents flow in the same winding, called the input winding, or primary; the output winding is then called the secondary.

It's also useful to classify electric machines in terms of their energy conversion characteristics. A machine acts as a generator if it converts mechanical energy from a prime mover. A machine is classified as a motor if it converts electrical energy to mechanical form. Electric motors are used to provide forces and torques to generate motion in countless industrial applications.

The characteristics of several common types of machines are summarized in Table 9. 1. 1. It's a useful tool for comparing the various types of motors. Also it will provide a convenient starting point for you when you face the problem of selecting the proper motor for one of your systems.

Table 9.1.1 Characteristics of Electrical Motors

		Type	Power Range(hp)	Rotor	Stator	Comments and Applications
AC motors	Three phase	Induction	1～5 000	Squirrel cage	Three-phase armature windings	Simple rugged construction; very common; fans, pumps
				Wound field		Adjustable speed using rotor resistance; cranes, hoists
		Synchronous	1～5	Permanent magnet		Precise speed; transport sheet materials
			1 000～50 000	DC field winding		Large constant loads; potential for power-factor correction
	Single phase	Induction	$\frac{1}{3}$～5	Squirrel cage	Main and auxiliary windings	Several types; split phase, capacitor start, capacitor run; simple and rugged; many household applications: fans, water pumps, refrigerators
		Synchronous	$\frac{1}{10}$ or less	Reluctance or hysteresis	Armature winding	Low torque, fixed speed; timing applications
DC motors	Wound field	Shunt connected	10～200	Armature winding	Field winding	Industrial applications, grinding, machine tools, hoists
		Series connected				High torque at low speed: dangerous if not loaded; drills, automotive starting motors, (universal motor used for singlephase AC has high power/weight ratio)
		Compound connected				Can be designed to tailor torque-speed characteristic; traction motors
	Permanent-magnel field		$\frac{1}{20}$～10	Armature winding	Permanent magnets	Servo applications, machine tools, computer peripherals, automotive fans, window motors

9.2 Basic construction

As shown in Figure 9. 2. 1, an electrical motor consists of a stationary part, or stator, and a rotor, which is the rotating part connected to a shaft that couples the machine to its mechanical load. The shaft and rotor are supported by bearings so that they can rotate freely.

Depending on the type of machine, either the stator or the rotor (or both) contains current-carrying conductors configured into coils. Slots are cut into the stator and rotor to contain the windings and their insulation. Currents in the windings set up magnetic fields and interact with fields to produce torque.

Usually, the stator and the rotor are made of iron to intensify the magnetic field. As in transformers, if the magnetic field alternates in direction through the iron with time, the iron must be laminated to avoid large power losses due to eddy currents (In certain parts of some machines, the field is steady and lamination is not necessary).

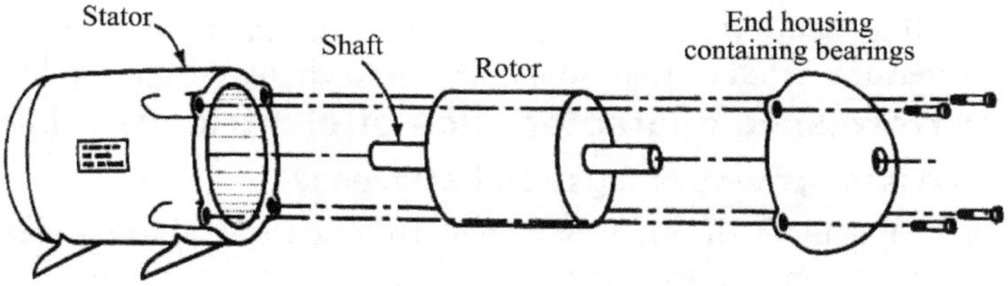

Figure 9. 2. 1 An electrical motor consists of a cylindrical rotor that spins inside a stator

9.3 Basic operation of electric machines

We have already learned how the magnetic field in electromechanical devices provides a form of coupling between electrical and mechanical systems. Intuitively, one can identify two aspects of this coupling, both of which play a role in the operation of electric machines:

(1) magnetic attraction and repulsion forces generate mechanical torque.

(2) the magnetic field can induce a voltage in the machine windings (coils) by virtue of Faraday's law.

Thus, we may think of the operation of an electric machine in terms of either a motor or a generator, depending on whether the input power is electrical and output power is mechanical (motor action), or the input power is mechanical and the output power is electrical (generator action). Figure 9. 3. 1 illustrates the two cases graphically.

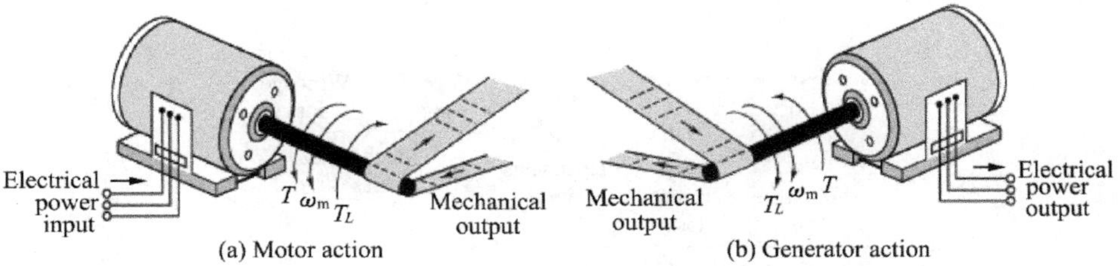

Figure 9. 3. 1 Generator and motor action in an electric machine

The coupling magnetic field performs a dual role, which may be explained as follows. When a current i flows through conductors placed in a magnetic field, a force is produced on each conductor. If these conductors are attached to a cylindrical structure, a torque is generated, and if the structure is free to rotate, then it will rotate at an angular velocity ωm. As the conductors rotate, however, they move through a magnetic field and cut through flux lines, thus, generating an electromotive force in opposition to the excitation. This emf is also called "counter" emf, as it opposes the source of the current i. If, on the other hand, the rotating element of the machine is driven by a prime mover (for example, an internal combustion engine), then an emf is generated across the coil that is rotating in the magnetic field (the armature). If a load is connected to the armature, a current i will flow to the load, and this current flow will in turn cause a reaction torque on the armature that opposes the torque imposed by the prime mover.

You see, then, that for energy conversion to take place, two elements are required:

(1) a coupling field, $\boldsymbol{B}$, usually generated in the field winding.

(2) an armature winding that supports the load current, i, and the emf, e.

9.4 Performance characteristics of electric machines

9.4.1 Losses, power ratings, and efficiency

Figure 9.4.1 depicts the flow of power from a three-phase electrical source through an induction motor to a mechanical load, such as a pump. Part of the electrical power is lost (converted to heat) due to resistance of the windings, hysteresis, and eddy currents. Similarly, some of the power that is converted to mechanical form is lost to friction and windage (i. e., moving the air surrounding the rotor and shaft) and part of the power loss to windage is sometimes intentional, because fan blades to promote cooling are fabricated as an integral part of the rotor.

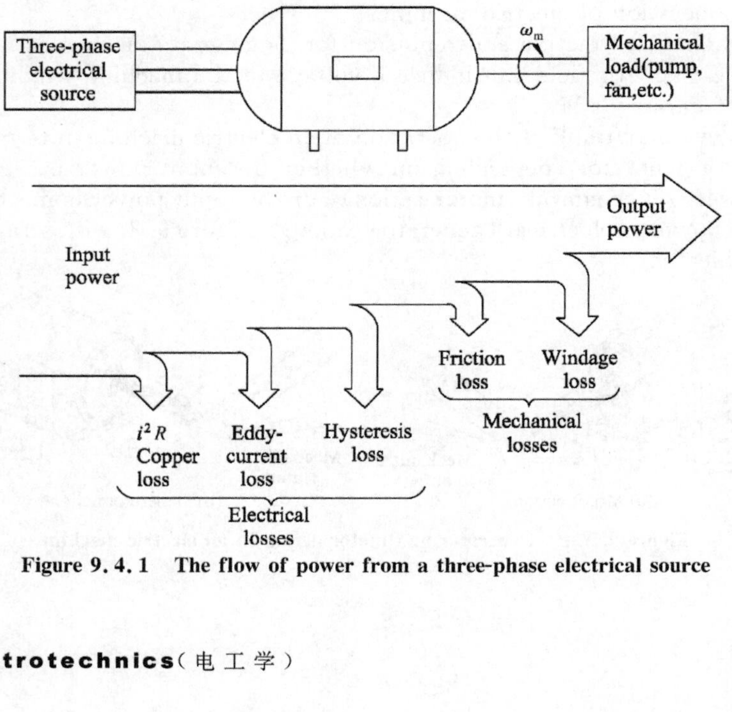

Figure 9.4.1 The flow of power from a three-phase electrical source

The electrical input power P_{in}, in watts, supplied by the three-phase source is given by

$$P_{in}=\sqrt{3}V_{rms}I_{rms}\cos(\theta) \tag{9.4.1}$$

where V_{rms} is the rms value of the line-to-line voltage, I_{rms} is the rms value of the line current, and $\cos(\theta)$ is the power factor.

The mechanical output power is

$$P_{out}=T_{out}\omega_m \tag{9.4.2}$$

in which P_{out} is the output power in watts, T_{out} is the output torque in newton-meters, and ω_m is the angular speed of the load in radians per second.

Also, torque may be given in foot-pounds instead of in newton-meters. The conversion relationship is

$$T_{\text{foot-pounds}}=T_{\text{newton-meters}}\times 0.7376$$

The power rating of a motor is the output power that the motor can safely produce on a continuous basis. It is important to realize that most motors can supply output power varying from zero to several times their rated power, depending on the mechanical load. It is up to the system designer to ensure that the motor is not overloaded.

The chief output power limitation of motors is their temperature rise due to losses. Thus, a brief overload that does not cause significant rise in temperature is often acceptable.

The power efficient of a motor is given by

$$\eta=\frac{P_{out}}{P_{in}}\times 100\%$$

Well-designed electrical motors operating close to their rated capacity have efficiency in the range of 85% to 95%. On the other hand, if the motor is called upon to produce only a small fraction of its rated power, its efficiency is generally much lower.

9.4.2 Torque-speed characteristics

Consider a system in which a three-phase induction motor drives a load such as a pump. Figure 9.4.2 shows the torque produced by the motor versus speed.

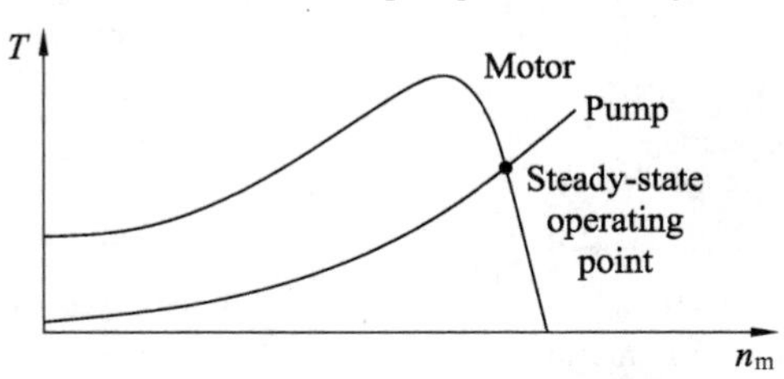

Figure 9.4.2 The torque-speed characteristics of an induction motor

The torque required to drive the load is also shown in Figure 9.4.2. Suppose that the system is at a standstill and then a switch is closed connecting the electrical source to the motor. At low speed, the torque produced by the motor is larger than that needed to drive the load. The excess torque causes the system to accelerate. Eventually, the speed stabilizes at the point for which the torque produced by the motor equals the torque needed to drive the load.

Now consider the torque-speed characteristics for a three-phase induction motor and a load consisting of a hoist shown in Figure 9.4.3. Here, the starting torque of the motor is less than that demanded by the load. Thus, if power is applied from a standing start, the system does not move. In this case, excessive currents are drawn by the motor, and unless fuses or other protection equipment disconnect the source, the motor could overheat and be destroyed.

Even though the motor cannot start the load shown in Figure 9.4.3, notice that the motor is capable of keeping the load moving once the speed exceeds n_1. Perhaps this could be accomplished with a mechanical clutch.

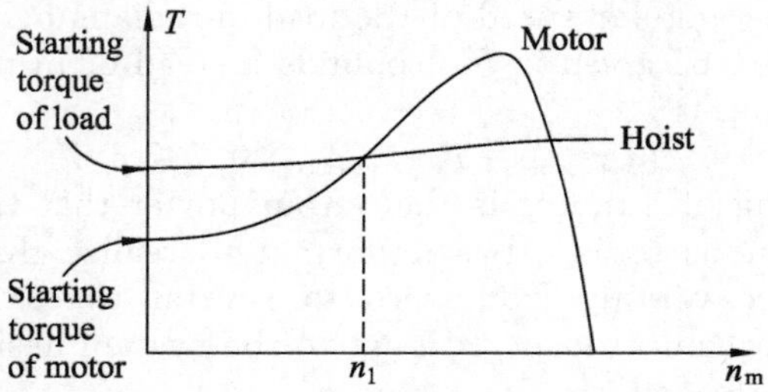

Figure 9.4.3 The torque-speed characteristics of a three-phase induction motor

The various types of motors have different torque-speed characteristics. Some examples are shown in Figure 9.4.4. It is important for the system designer to choose a motor suitable for the load requirements.

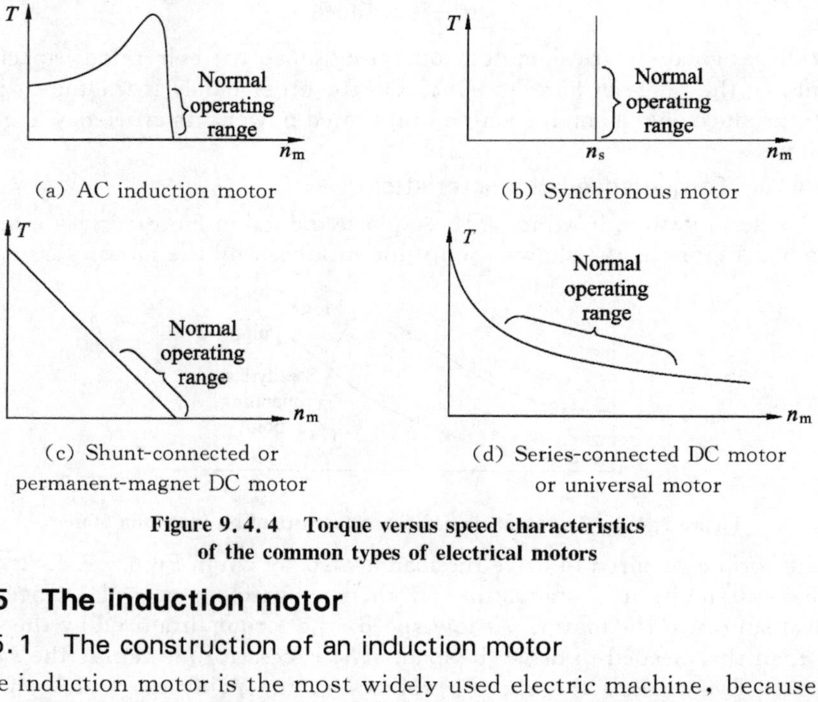

Figure 9.4.4 Torque versus speed characteristics of the common types of electrical motors

9.5 The induction motor

9.5.1 The construction of an induction motor

The induction motor is the most widely used electric machine, because of its relative simplicity of construction. The primary advantage of the induction machine, which is almost exclusively used as a motor is that no separate excitation is required for the rotor. The rotor typically consists of one of two arrangements: **a squirrel cage**, or a **wound rotor.** The former contains conducting bars short-circuited at the end and embedded within it; the latter consists of a multiphase winding similar to that used for the stator, but electrically short-circuited. In either case, the induction motor operates by virtue of currents induced from the stator field in the rotor. In this respect, its operation is similar to that of a transformer, in that currents in the stator (which acts as a primary coil) induce currents in the rotor (acting as a secondary coil). In most induction motors, no external electrical connection is required for the rotor, thus permitting a simple, rugged construction, without the need for slip rings or brushes. The induction motor does not operate at synchronous speed, but at a somewhat lower speed, which is dependent on the load. Figure 9.5.1 illustrates the construction of a squirrel-cage induction motor. Figure 9.5.2 illustrates the appearance of a squirrel-cage rotor. The following discussion will focus mainly on this very common configuration.

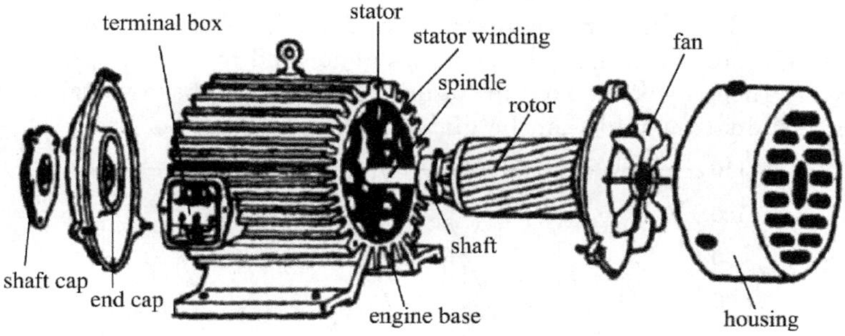

Figure 9.5.1 A squirrel-cage induction motor

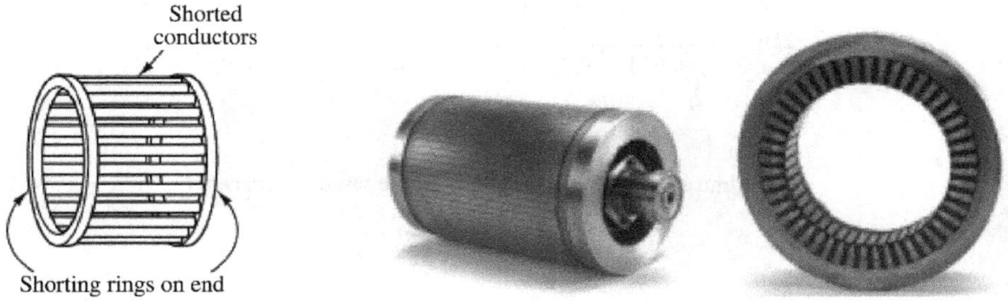

(a) Conductors in a squirrel-cage rotor (b) Rotor (c) Cross section of stator

Figure 9.5.2 The appearance of a squirrel-cage rotor

9.5.2 Rotating magnetic fields

As mentioned previously, the fundamental principle of operation of AC machines is the generation of a rotating magnetic field, which causes the rotor to turn at a speed that depends on the speed of rotation of the magnetic field. We shall now explain how a rotating magnetic field can be generated in the stator and air gap of an induction motor by means of AC currents.

Figure 9.5.3 is a cross-sectional view of a induction motor. In the Figure, a box with a cross inscribed in it indicates current flowing into the page, while a dot represents current out of the plane of the page.

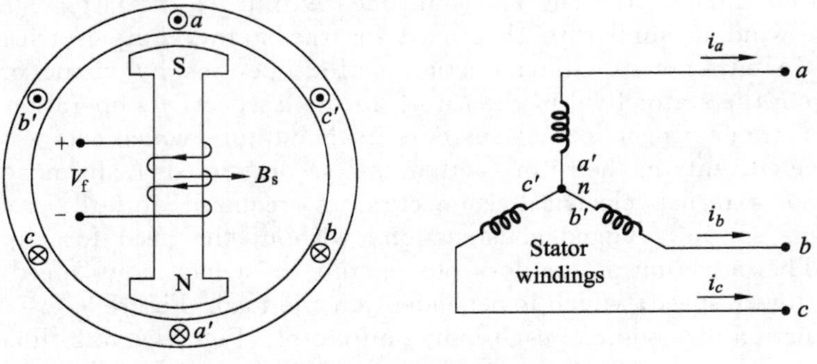

Figure 9. 5. 3 Two-pole three-phase stator

Consider the stator shown in Figure 9. 5. 3, which supports windings $a-a'$, $b-b'$ and $c-c'$. The coils are geometrically spaced 120° apart, and a three-phase voltage is applied to the coils. The currents generated by a three-phase source are also spaced by 120°, as illustrated in Figure 9. 5. 4. The phase voltages referenced the neutral terminal, would then be given by the expressions

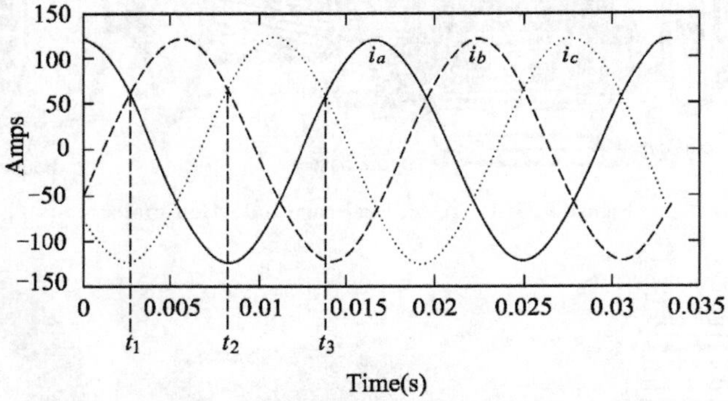

Figure 9. 5. 4 Three-phase stator winding currents

$$v_a = A\cos(\omega t)$$
$$v_b = A\cos(\omega t - 120°)$$
$$v_c = A\cos(\omega t + 120°)$$

where ω is the frequency of the AC supply, or line frequency. The coils in each winding are arranged in such a way that the flux distribution generated by any one winding is approximately sinusoidal. Such a flux distribution may be obtained by appropriately arranging groups of coils for each winding over the stator surface. Since the coils are spaced 120° apart, the flux distribution resulting from the sum of the contributions of the three windings is the sum of the fluxes due to the separate windings, as shown in Figure 9. 5. 5. Thus, the flux in a three-phase machine rotates in space according to the vector diagram of Figure 9. 5. 6, and is constant in amplitude.

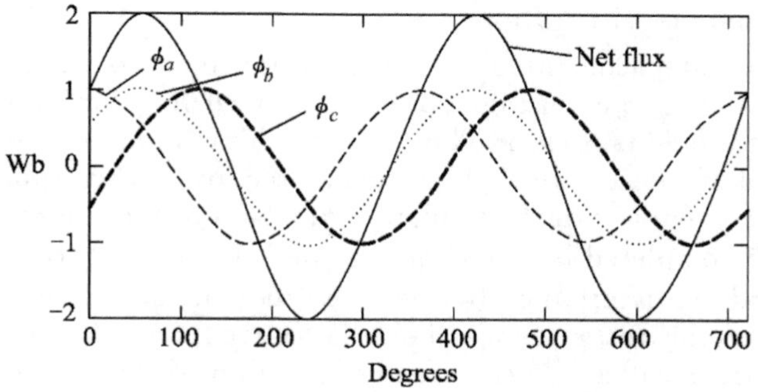

Figure 9.5.5 Flux distribution in three-phase

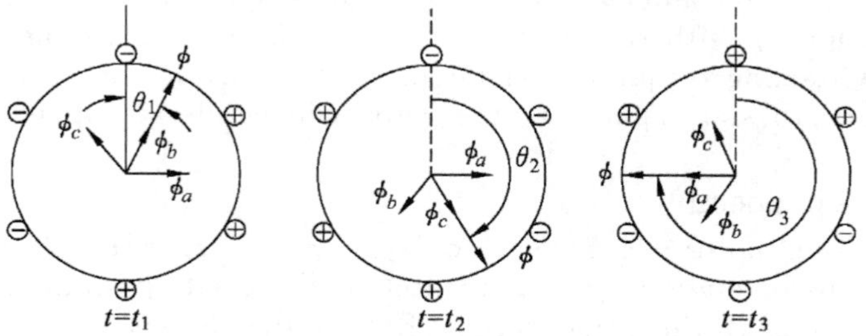

Figure 9.5.6 Rotating flux in a three-phase machine

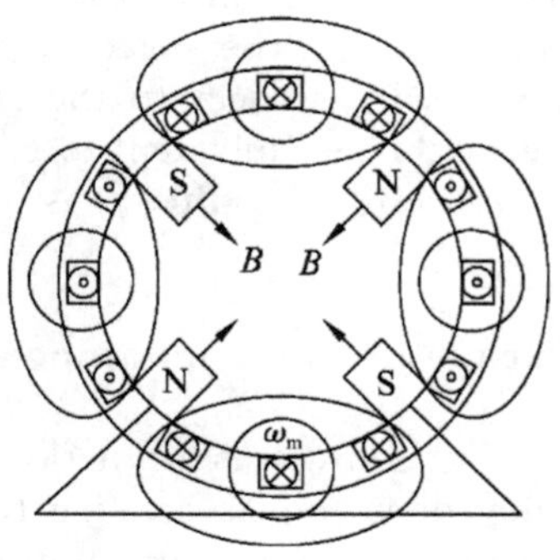

Figure 9.5.7 Four-pole stator

Since the resultant flux of Figure 9.5.5 is generated by the currents of Figure 9.5.4, the speed of rotation of the flux must be related to the frequency of the sinusoidal phase currents. In the case of the stator of Figure 9.5.3, the number of magnetic poles resulting from the winding configuration is two; however, it is also possible to configure the windings so that they have more poles. For example, Figure 9.5.7 depicts a simplified view of a four-pole stator.

In general, the speed of the rotating magnetic field is determined by the frequency of the excitation current, f, and by the number of poles present in the stator, p, according to the equation

$$n_s = \frac{120f}{p} \tag{9.5.1}$$

where n_s is usually called the synchronous speed.

Now, the structure of the windings in the preceding discussion is the same whether the AC machine is a motor or a generator; the distinction between the two depends on the direction of power flow. In a generator, the electromagnetic torque is a reaction torque that opposes rotation of the machine; this is the torque against which the prime mover does work. In a motor, on the other hand, the rotational (motional) voltage generated in the armature opposes the applied voltage; this voltage is the counter (or back) emf.

9.5.3 Principle of rotation

By now, we are acquainted with the notion of a rotating stator magnetic field. Imagine that a squirrel-cage rotor is inserted in a stator in which such a rotating magnetic field is present. The stator field will induce voltages in the cage conductors, and if the stator field is generated by a three-phase source, the resulting rotor currents—which circulate in the bars of the squirrel cage, with the conducting path completed by the shorting rings at the end of the cage—are also three-phase, and are determined by the magnitude of the induced voltages and by the impedance of the rotor. Since the rotor currents are induced by the stator field, the number of poles and the speed of rotation of the induced magnetic field are the same as those of the stator field, if the rotor is at rest. Thus, when a stator field is initially applied, the rotor field is synchronous with it, and the fields are stationary with respect to each other. Thus, according to the earlier discussion, a starting torque is generated. If the starting torque is sufficient to cause the rotor to start spinning, the rotor will accelerate up to its operating speed.

9.5.4 Slip speed

An induction motor can never reach synchronous speed; if it did, the rotor would appear to be stationary with respect to the rotating stator field, since it would be rotating at the same speed. But in the absence of relative motion between the stator and rotor fields, no voltage would be induced in the rotor. Thus, an induction motor is limited to speeds somewhere below the synchronous speed, n_s. Let the speed of rotation of the rotor be n; then, the rotor is losing ground with respect to the rotation of the stator field at a speed $(n_s - n)$. In effect, this is equivalent to backward motion of the rotor at the **slip speed**, defined by $(n_s - n)$. The **slip**, s, is usually defined as a fraction of n_s

$$s = \frac{n_s - n}{n_s} \tag{9.5.2}$$

which leads to the following expression for the rotor speed:

$$n = n_s(1 - s) \tag{9.5.3}$$

The slip, s, is a function of the load, and the amount of slip in a given motor is dependent on its construction and rotor type (squirrel cage or wound rotor). Since there is a relative motion between the stator and rotor fields, voltages will be induced in the rotor at a frequency called the **slip frequency**, related to the relative speed of the two fields. This gives rise to an interesting phenomenon: the rotor field travels relative to the rotor at the slip speed sn_s, but the rotor is mechanically traveling at the speed $(1-s)n_s$, so that the net effect is that the rotor field travels at the speed

$$sn_s + (1 - s)n_s = n_s \tag{9.5.4}$$

that is, at synchronous speed. The fact that the rotor field rotates at synchronous speed—although the rotor itself does not—is extremely important, because it means that the stator and rotor fields will continue to be stationary with respect to each other, and therefore a net torque can be produced.

Example 9.5.1 Induction motor analysis

Find the full load rotor slip and frequency of the induced voltage at rated speed in a four-pole induction motor. Motor ratings: 230 V; 60 Hz; full-load speed: 1 725 rev/min.

Solution: The synchronous speed of the motor is

$$n_s=\frac{120f}{p}=\frac{120\times 60}{4}=1\ 800\ \text{rev/min}$$

The slip is
$$s=\frac{n_s-n}{n_s}=\frac{1\ 800-1\ 725}{1\ 800}0.041\ 7$$

The rotor frequency is

$$f_R=sf=0.041\ 7\times 60=2.5\ \text{Hz}$$

9.5.5 Equivalent circuit of an induction motor

The induction motor can be described by means of an equivalent circuit, which is essentially that of a rotating transformer. Figure 9.5.8 depicts such a circuit model, where:

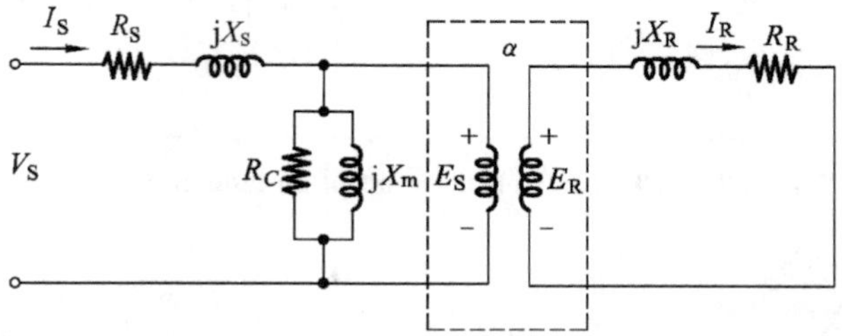

Figure 9.5.8 Circuit model for induction machine

R_S=stator resistance per phase, R_R=rotor resistance per phase, X_S=stator reactance per phase, X_R=rotor reactance per phase, X_m=magnetizing (mutual) reactance, R_C=equivalent core-loss resistance, E_S=per-phase induced voltage in stator windings, and E_R=per-phase induced voltage in rotor windings.

The primary internal stator voltage, E_S, is coupled to the secondary rotor voltage, E_R, by an ideal transformer with an effective turns ratio α. For the rotor circuit, the induced voltage at any slip will be

$$E_R=sE_{R0} \tag{9.5.5}$$

where E_{R0} is the induced rotor voltage at the condition in which the rotor is stationary. Also, $X_R=\omega_R L_R=2\pi f_R L_R=2\pi s f L_R=sX_{R0}$, where $X_{R0}=2\pi f L_R$ is the reactance when the rotor is stationary. The rotor current is given by the expression:

$$I_R=\frac{E_R}{R_R+jX_R}=\frac{sE_{R0}}{R_R+jsX_{R0}}=\frac{E_{R0}}{\dfrac{R_R}{s}+jX_{R0}} \tag{9.5.6}$$

The resulting rotor equivalent circuit is shown in Figure 9.5.9.

The voltages, currents, and impedances on the secondary (rotor) side can be reflected to the primary (stator) by means of the effective turns ratio. When this transformation is effected, the transformed rotor voltage is given by

$$E_2 = E'_R = \alpha E_{R0} \quad (9.5.7)$$

The transformed (reflected) rotor current is

$$I_2 = \frac{I_R}{\alpha} \quad (9.5.8)$$

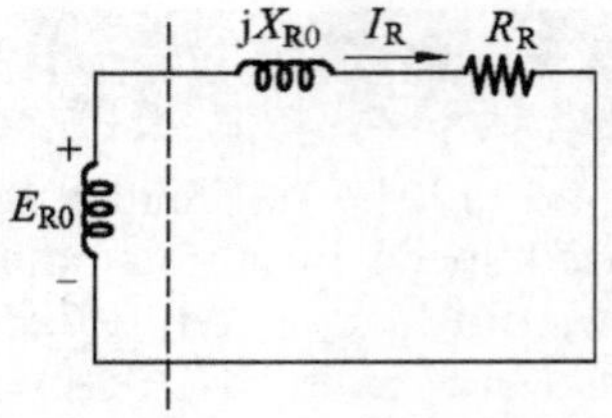

Figure 9.5.9 Rotor equivalent circuit

The transformed rotor resistance can be defined as

$$R_2 = \alpha^2 R_R \quad (9.5.9)$$

and the transformed rotor reactance can be defined by

$$X_2 = \alpha^2 X_{R0} \quad (9.5.10)$$

The final per-phase equivalent circuit of the induction motor is shown in Figure 9.5.10.

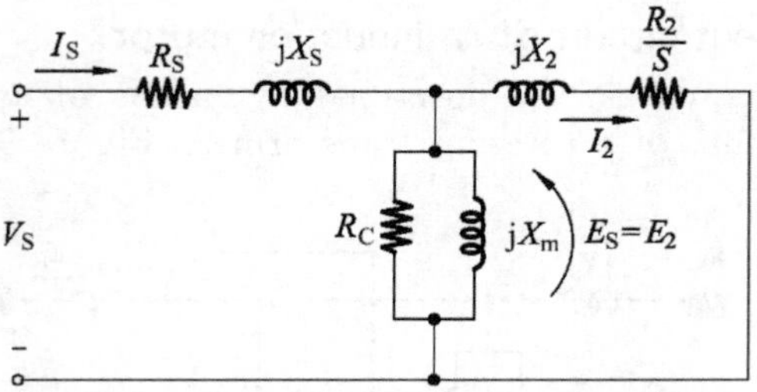

Figure 9.5.10 Equivalent circuit of an induction machine

Example 9.5.2

Determine the following quantities for an induction motor using the circuit models of Figures 9.5.9 to 9.5.10.

(a) Speed;

(b) Stator current;

(c) Power factor;

(d) Output torque.

Given Data: Motor ratings—460 V; 50 Hz; four poles; $s = 0.022$; $P = 14$ hp; $R_S = 0.641$; $R_2 = 0.332$; $X_S = 1.106$; $X_2 = 0.464$; $X_m = 26.3$.

Assumptions: Use per-phase analysis. Neglect core losses ($R_C = 0$).

Solution: (a) The per-phase equivalent circuit is shown in Figure 9.5.10. The synchronous speed is found to be

$$n_s = \frac{120f}{p} = \frac{120 \times 50}{4} = 1\,500 \text{ rev/min}$$

$$\omega_s = 1\,500 \times \frac{2\pi}{60} = 157 \text{ rad/s}$$

The rotor mechanical speed is

$$n=(1\text{-}s)n_s=1\ 467\ \text{rev/min}$$
$$\omega_m=(1\text{-}s)\omega_s=153.5\ \text{rad/s}$$

(b) The reflected rotor impedance is found from the parameters of the per-phase circuit to be

$$Z_2=\frac{R_2}{s}+\text{j}X_2=\frac{0.332}{0.022}+\text{j}0.464=15.09+\text{j}0.464\ \Omega$$

The combined magnetization plus rotor impedance is therefore equal to

$$Z=\frac{1}{\dfrac{1}{\text{j}X_m}+\dfrac{1}{Z_2}}=\frac{1}{-\text{j}0.038+0.066\ 2\angle-1.76^\circ}=12.94\angle31.1^\circ\ \Omega$$

and the total impedance is

$$Z_{\text{total}}=Z_S+Z=0.641+\text{j}1.106+11.8+\text{j}6.68$$
$$=11.72+\text{j}7.79=14.07\angle33.6^\circ\ \Omega$$

Finally, the stator current is given by

$$I_s=\frac{V_s}{Z_{\text{total}}}=\frac{\dfrac{460\angle0^\circ}{\sqrt{3}}}{14.07\angle33.6^\circ}=18.88\angle\text{-}33.6^\circ\text{A}$$

(c) The power factor is

$$pf=\cos33.6^\circ=0.883\quad \text{lagging}$$

(d) The output power is

$$P_{\text{out}}=14\times746=10.444\ \text{kW}$$

and the output torque is

$$T=\frac{P_{\text{out}}}{\omega_m}=\frac{10.444}{153.5}\times10^3=68.04\ \text{N}\cdot\text{m}$$

9.5.6 Performance of induction motors

The performance of induction motors can be described by torque-speed curves similar to those already used for DC motors. Figure 9.5.11 depicts an induction motor torque-speed curve, with five torque ratings marked *a* through *e*. Point *a* is the starting torque, also called **breakaway torque**, and is the torque available with the rotor "locked", that is, in a stationary position. Under this condition, the frequency of the voltage induced in the rotor is the highest, since it is equal to the frequency of rotation of the stator field; consequently, the inductive reactance of the rotor is the greatest. As the rotor accelerates, the torque drops off, reaching a maximum value called the **pull-up torque** (point *b*); this typically occurs somewhere between 25% ~ 40% of synchronous speed. As the rotor speed continues to increase, the rotor reactance decreases further (since the frequency of the induced voltage is determined by the relative speed of rotation of the rotor with respect to the stator field). The torque becomes a maximum when the rotor inductive reactance is equal to the rotor resistance; maximum torque is also called **breakdown torque** (point *c*). Beyond this point, the torque drops off, until it is zero at synchronous speed, as discussed earlier. Also marked on the curve are the 150 percent torque (point *d*), and the rated torque (point *e*).

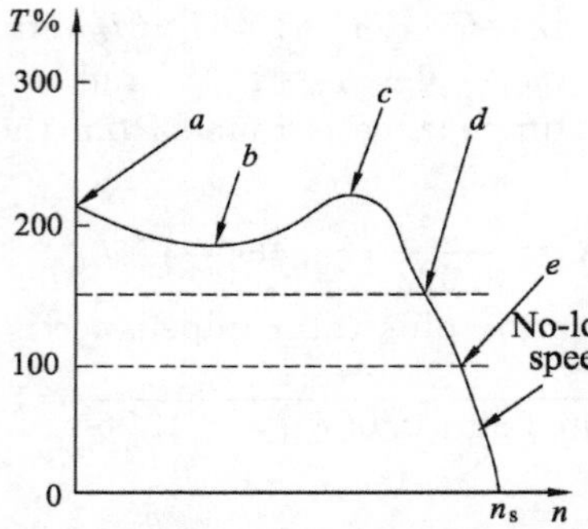

Figure 9. 5. 11 Performance curve for induction motor

A general formula for the computation of the induction motor steady-state torque-speed characteristic is

$$T=\frac{mV_s^2R_R/s}{\omega_e\left[\left(R_S+\frac{R_R}{s}^2\right)+(X_S+X_R)^2\right.} \tag{9.5.11}$$

where m is the number of phases.

9.5.7 AC motor speed and torque control

As explained in an earlier section, AC machines are constrained to fixed-speed or near fixed-speed operation when supplied by a constant-frequency source. Several simple methods exist to provide limited speed control in an AC induction machines; more complex methods, involving the use of advanced power electronics circuits can be used if the intended application requires wide-bandwidth control of motor speed or torque. In this subsection, we provide a general overview of available solutions.

9. 5. 7. 1 Pole number control

The (conceptually) easiest method to implement speed control in an induction machine is by varying the number of poles. Equation 9. 1. 1 explains the dependence of synchronous speed in an AC machine on the supply frequency and on the number of poles. For machines operated at 60 Hz, the speeds in Table 9. 5. 1 can be achieved by varying the number of magnetic poles in the stator winding.

Table 9. 5. 1

Number of poles	2	4	6	8	12
n, rev/min	3 600	1 800	1 200	800	600

Motor stators can be wound so that the number of pole pairs in the stators can be varied by switching between possible winding connections. Such switching requires that care be taken in timing it to avoid damage to the machine.

9. 5. 7. 2 Slip control

Since the rotor speed is inherently dependent on the slip, slip control is a valid means of achieving some speed variation in an induction machine. Since motor torque falls with the square of the voltage (see Equation (9. 5. 11)), it is possible to change the slip by changing the motor torque through a reduction in motor voltage.

This procedure allows speed control over the range of speeds that allow stable motor operation. With reference to Figure 9.5.11, this is possible only above point *c*, that is, above the breakdown torque.

9.5.7.3 Rotor control

For motors with wound rotors, it is possible to connect the rotor slip rings to resistors; adding resistance to the rotor increases the losses in the rotor, and therefore causes the rotor speed to decrease. This method is also limited to operation above the breakdown torque though it should be noted that the shape of the motor torque-speed characteristic changes when the rotor resistance is changed.

9.5.7.4 Frequency regulation

The last two methods cause additional losses to be introduced in the machine. If a variable-frequency supply is used, motor speed can be controlled without any additional losses. As seen in Equation (9.5.1), the motor speed is directly dependent on the supply frequency, as the supply frequency determines the speed of the rotating magnetic field. However, to maintain the same motor torque characteristics over a range of speeds, the motor voltage must change with frequency, to maintain a constant torque. Thus, generally, the V/Hz ratio should be held constant.

This condition is difficult to achieve at start-up and at very low frequencies, in which cases the voltage must be raised above the constant V/Hz ratio that will be appropriate at higher frequency.

Summary

1. The main mechanical components of an electric machine are the stator, rotor, and air gap.

Electrically, the important parameters are the armature (load current-carrying) circuit, and the field (magnetizing) circuit. Magnetic fields establish the coupling between the electrical and mechanical systems. Typically, DC machines have the armature winding on the rotor, while AC machines have it on the stator.

2. Electric machines are described in terms of their mechanical characteristics, their torque-speed curves, and their electrical characteristics, including current and voltage requirements. Losses and efficiency are important parts of the operation of electric machines, and it should be recognized that there will be electrical losses (due to the resistance of the windings), mechanical losses (friction and windage), and magnetic core losses (eddy currents, hysteresis).

3. Induction machines (of the squirrel-cage type) do not require a field excitation, since this is provided by electromagnetic induction.

4. The performance of electric machines can be approximately predicted with the use of circuit models, or of performance curves. The selection of a particular machine for a given application is driven by many factors, including the availability of suitable electrical supplies (or prime movers), the type of load, and various other concerns, of which heat dissipation and thermal characteristics are probably the most important.

Problems

9.1 The percent speed regulation of a motor is 10%. If the full-load speed is 50π rad/s, find (a) the no-load speed in rad/s, and (b) the no-load speed in rev/min.

9.2 The percent voltage regulation for a 250 V generator is 10%. Find the no-load voltage of the generator.

9.3 The nameplate of a three-phase induction motor indicates the following values:

H. P. = 10; V = 220 V; R. P. M. = 1 750; Service factor = 1.15; Temperature rise = 60 ℃; Amp = 30 A. Find the rated torque, rated volt-amperes, and maximum continuous output power.

9.4 A three-phase induction motor has six poles. (a) If the line frequency is 60 Hz, calculate the speed of the magnetic field in rev/min. (b) Repeat the calculation if the frequency is changed to 50 Hz.

9.5 A four-pole induction motor operating at a frequency of 60 Hz has a full-load slip of 4%. Find the frequency of the voltage induced in the rotor (a) at the instant of starting and (b) at full load.

9.6 A four-pole, 1 746 rev/min, 220 V, 3-phase, 60 Hz, 10 hp, Y-connected induction machine has the following parameters: $R_S=0.4$, $R_2=0.14$, $X_m=16$, $X_S=0.35$, $X_2=0.35$, $R_C=0$. Using Figure 16.39, find: (a) the stator current; (b) the rotor current; (c) the motor power factor; and (d) the total stator power input.

9.7 A three-phase, 220 V, 50 Hz induction motor runs at 1 440 rev/min. Determine:

(a) the number of poles (for minimum slip); (b) the slip; (c) and the frequency of the rotor currents.

9.8 An eight-pole, three-phase, 220 V, 60 Hz, induction motor has the following model impedances: $R_S=0.78$; $X_s=0.56$; $X_m=32$; $R_R=0.28$; $X_R=0.84$. Find the input current and power factor of this motor for $s=0.02$.